International Trends in Manufacturing Technology

ROBOT VISION

Edited by Professor Alan Pugh, University of Hull, UK

IFS (Publications) Ltd., UK

Springer-Verlag

Berlin · Heidelberg · New York · Tokyo

1983

British Library Cataloguing Publication Data

Robot Vision
1. Information display systems
i. Pugh, Alan
621.38'0414   TK7882.16

ISBN 0-903608-32-4 IFS (Publications) Ltd.
ISBN 3-540-12073-4 Springer-Verlag · Berlin · Heidelberg · New York · Tokyo
ISBN 0-387-12073-4 New York · Heidelberg · Berlin · Tokyo

Typesetting by Fotographics (Bedford) Ltd., UK
Printed and bound in Great Britain by
Anchor Brendon Ltd, Tiptree, Essex, UK

Acknowledgements

IFS (Publications) Limited wishes to express its acknowledgement and appreciation to the following publishers/organisations for supplying the articles reprinted within its contents.

Japan Industrial Robot Association,
c/o Kikaishinko Bldg.,
3-5-8 Shibakoen, Minato-ku,
Tokyo, Japan.

Hitachi Ltd.,
Nippon Building,
6-2, 2-chome, Ohtemachi,
Chiyoda-ku, Tokyo 100, Japan.

Crane, Russak & Co. Inc.,
3 East 44th Street, New York,
N.Y. 10017, U.S.A.

Science and Engineering Research
Council,
Swindon, Wilts, U.K.

The MIT Press,
28 Carleton Street,
Cambridge, Massachusetts 02142,
U.S.A.

Carl Hanser GmbH & Co.
"Feinwerktechnik & Messtechnik",
Kolbergerstrasse 22,
8000 Munchen 80,
Germany.

Society of Manufacturing Engineers,
One SME Drive, PO Box 930,
Dearborn, Michigan 48128,
U.S.A.

The Institution of Electrical Engineers,
Savoy Place,
London WC2R 0BL, U.K.

Institute of Electrical & Electronic
Engineers,
345 East 47 Street,
New York, N.Y. 10017,
U.S.A.

About this Series

International Trends and Manufacturing Technology

The advent of microprocessor controls and robotics is rapidly changing the face of manufacturing throughout the world. Large and small companies alike are adopting these new methods to improve the efficiency of their operations. Researchers are constantly probing to provide even more advanced technologies suitable for application to manufacturing. In response to these advances IFS (Publications) Ltd., is to publish a new series of books on topics that will highlight the developments that are taking place in manufacturing technology. The series aims to be informative and educational.

Subjects to be covered in the new series will include:

Robot vision
Visual inspection
Flexible assembly automation
Automated welding systems
Flexible manufacturing systems
Lasers in manufacturing

The series is intended for manufacturing managers, production engineers and those working on research into advanced manufacturing methods. Each book will be published in hard cover and will be edited by a specialist in the particular field.

The first in the series – Robot Vision – is under the editorship of Professor Alan Pugh, Head of Department of Electronic Engineering, University of Hull, England, UK. The series editors are: Jack Hollingum, John Mortimer, Brian Rooks.

Finally, I express my gratitude to the authors whose works appear in this publication.

John Mortimer,
Managing Director,
IFS (Publications) Limited

PREFACE

Over the past five years robot vision has emerged as a subject area with its own identity. A text based on the proceedings of the Symposium on Computer Vision and Sensor-based Robots held at the General Motors Research Laboratories, Warren, Michigan in 1978, was published by Plenum Press in 1979. This book, edited by George G. Dodd and Lothar Rossol, probably represented the first identifiable book covering some aspects of robot vision.

The subject of robot vision and sensory controls (RoViSeC) occupied an entire international conference held in the Hilton Hotel in Stratford, England in May 1981. This was followed by a second RoViSeC held in Stuttgart, Germany in November 1982. The large attendance at the Stratford conference and the obvious interest in the subject of robot vision at international robot meetings, provides the stimulus for this current collection of papers.

Users and researchers entering the field of robot vision for the first time will encounter a bewildering array of publications on all aspects of computer vision of which robot vision forms a part. It is the grey area dividing the different aspects of computer vision which is not easy to identify. Even those involved in research sometimes find difficulty in separating the essential differences between vision for automated inspection and vision for robot applications. Both of these are to some extent applications of pattern recognition with the underlying philosophy of each defining the techniques used.

The contributors to this book have been identified from those in our community who have demonstrated a track record in robot vision and the papers, most of which have been published previously elsewhere, have been selected as the most recent representative contributions from these authors. The aim of the book, therefore, is to bring together the most significant work of those active in all aspects of robot vision to form a nucleus of knowledge representing the current 'state of the art'.

The first two papers give introductory reviews, which are intended to support the contributions which follow. Subsequent papers cover aspects of image processing for robot applications through to a description of commercial vision systems available today. It is the realisation of suitable vision sensors, coupled with the illumination of the work area and supported by elegant software which forms the key to successful robot vision. Perhaps the most active area of application which illustrates this point, arises under the heading of weld guidance, where the promise of robot vision and elementary structured light is most evident. Robot vision must play its part in sensor guided assembly to complement the work on tactile wrists and grippers. In the act of component retrieval, it is robot vision which will dominate as the primary sense. Readers will note that the commercially available robot systems are designed primarily for high speed scene analysis and will not always accommodate the more precise requirements for automated inspection. It is important to recognise the difference in application area to understand the important properties of these commercial vision systems.

After digesting the contents of this book, readers will be left in no doubt that robot vision and its application to robot technology is in its infancy. Those of us who have been involved in the area for more than a decade, find that the successful application of robot vision is tempered by the problem areas which the technology is itself identifying. It is an exciting field with enormous challenge. Only with relentless pursuit will second generation robotics become an industrial reality by incorporating robot vision and other sensory methods as an essential part of the specification.

The editor wishes to express his appreciation to all who have contributed to this book. A particular vote of thanks is offered to Dr. B. W. Rooks who looked after the detailed negotiations with authors and the publisher.

Alan Pugh,
Hull

September 1982

CONTENTS

Chapter 1:
Reviews

Chapter 2:
Visual Processing Techniques

Chapter 3:
Research

Chapter 4:
Developments – Weld Guidance

Chapter 5:
Developments – Assembly / Part
Presentation

Chapter 6:
Applications

Chapter 7:
Commercial Robot Vision Systems

Chapter 1
REVIEWS

The development of vision systems for robots was started over 10 years ago. The two papers review this development and the industrial systems now emerging.

SECOND GENERATION ROBOTICS

Alan Pugh, Department of Electronic Engineering, University of Hull, UK

Presented at the 12th International Symposium on Industrial Robots, 9-11 June 1982, Paris, France

Introduction

During the early years of their existence, industrial robots represented a solution in search of a problem. At the first International Symposium on Industrial Robots held in Chicago in 1970[1] the delegates were treated to a brief catalogue of robot applications in industries which were invariably hot, smelly and involved jobs requiring a great deal of muscle power. This was the era when the industrial robot was a mere curiosity and its existence was known only to a relatively few informed industrialists. It was the combination of the industrial robot with the problem of spot welding automobile bodies which allowed the versatility of the industrial robot to be properly exploited. This single application transformed the industrial robot scene overnight which resulted in an escalation of the number of robots employed in industry coupled with a liberal coverage in the media which in turn stimulated interest to extend the application of industrial robots to other industrial tasks.

Despite the increasing use of robots in industries having a hostile environment – for example diecasting and forging – the greater success in application has hitherto been in areas where precise contact with the work place has not been an essential ingredient of the application.[2] In addition to spot welding which falls into this category, paint spraying has been a popular and successful application and considerable success in recent years has been established with seam welding. While the industrial robot was originally conceived with a point-to-point mode of operation, it has been the continuous path mode of operation which has dominated most industrial robot applications. The situation existing now is that robots have become a natural and essential feature for the establishment of a manufacturing process particularly if this is being assembled on a green-field factory site. No longer do potential users need to wait for some adventurous industrialist to experiment with applications for the first time; most of this has been done and applications in the *non-sensory* handling and processing areas are well catalogued. Further, developments in robot structures and mechanisms now give the user a superb range of devices/products from which to choose as any visit to an exhibition of robots will demonstrate.

Where is 'generation 2'?

From the very beginning of industrial robot technology, the prospect of intelligent control coupled to environmental sensors has been 'just around the corner'. That 'corner' is now over 10 years old and still no really satisfactory realisation of the 'second generation' has been forthcoming. The impressive and promising demonstrations of 'artificial intelligence' at the First International Symposium[1] gave way to more relevant realisations of sensory interaction during the 1970s and a few commercial vision systems are now available.[3, 4, 5] These however, represent the product of research of about a decade ago and demand a high contrast image for reliable operation.

Cost too plays its part as a deterrent in the sluggish approach of the 'second generation'. We see again the natural reluctance of industrial users not wishing to be first in the field. Acknowledging that substantial developments might exist behind the 'closed doors' of industrial confidentiality, only a few brave experiments towards the 'second generation' have appeared in the technical press. Incentive plays its part in the use of sensory control for certain applications and the microelectronic industry has provided a need for such radical thinking. Examples of this work have been published for some time.[6, 7] The need for vision in this industry is associated with the high degree of visual feedback required during the act of device fabrication and assembly and the substantial geometric content of integrated circuit pellets has encouraged the application of machine vision to the problem of automatic alignment and wire bonding. The important aspect of this application which has demonstrated reliable operation must be the relative ease of control of the workplace and the illumination of the assembly area.

Can this success in the use of sensory control be regarded as a realisation of the 'generation 2' robot? Some would argue that it does but most would not accept the specialist mechanised handling as representative of versatile 'programmable automation'. Published attempts of sensory control of robots for shop-floor application are difficult to find. Single experimental applications of both visual and tactile feedback have been implemented although we have an example in the United Kingdom of a company marketing as a standard feature a vision controller for paint spraying[8] which is supported by several years of operating experience.

The application areas for the first generation robots are rarely associated with automated assembly excepting the applications of non-programmable placement devices (or robots). The breakthrough required to extend the application of programmable robots into automated assembly which will open the door to flexible manufacturing systems can only be realised with sensory control to support the assembly process. Despite recent developments in robot architecture,[9] no programmable robot offers the kind of positioning precision to permit reliable assembly to take place without innovations of sensory control of the simplicity of the 'remote centre compliance'.[10] Automated programmable assembly coupled with sensory interactive handling represents the goal which best defines 'generation 2' robotics. The manufacturing areas which involve small to medium sized batches dominate in the industrial scene and those are the areas which are starved of automated solutions. To achieve success in this sector, a great deal of work is still required before the flexibility offered by highly expensive 'generation 2' devices can be justified.

Vision research

It is perhaps vision more than the senses of touch and hearing which has attracted the greatest research effort. However, robot vision is frequently confused with vision applied to automated inspection and even the artificial intelligence aspects of scene analysis. If a uniformed comparison is made between the technology of picture processing and the requirements of robot vision it is not possible to reconcile the apparent divide which exists between the two.

The essential requirements for success in robot vision might be summarised as follows:

- low cost,
- reliable operation,
- fundamental simplicity,
- fast image processing,
- ease of scene illumination.

These requirements are often diametrically opposed to the results of research effort published by research organisations. The processing of grey-scale images at high resolution often provides impressive results but inevitably this is achieved at the expense of processor architecture and processing time. Dedicated image processing systems[11] will attack the problem of processing speed in a most impressive way but there is often a desire on the part of many researchers to identify an area of technological challenge in image processing to satisfy their own research motivation rather than attempt a simplification of the imaging problems.

Probably the single aspect which causes difficulty but often overlooked is the control of illumination of the work area. This problem has been attacked by some researchers using 'planes of light'[12, 13] which might be regarded as a primitive application of structured lighting i.e. super-imposing on the work area a geometric pattern of light which is distorted by the workpieces. The success of this approach is manifested in the simplicity of binary image processing and a reduction in the magnitude of visual data to be analysed.[14] Developments of early demonstrations of robot vision using back lighting of the work area have reached a stage of restricted industrial application.[15] However, a feature of sensory techniques which have industrial potential, is that they are often application specific and cannot be applied generally.

The experiments linking image processing of televised images with robot applications, for example,[16] are in their infancy at present; the cost is high and the reliability of image processing is unlikely to be satisfactory for some time to come.

Image resolution

Sensing for robot applications is not dependent on a relentless pursuit for devices with higher and higher resolution. The fundamental consideration must be in the selection of *optimum* resolution for the task to be executed. There is a tendency to assume that the higher the resolution then the greater is the application range for the system. At this point in our evolution, we are not exploiting the 'state of the art' as much as we should. Considerable success has been achieved using a resolution as low as 50×50.[6] With serial processing architectures, this resolution will generate quite sufficient grey-scale data to test the ingenuity of image processing algorithms! Should processing time approach about 0.5 seconds, this will be noticeable in a robot associated with handling. However, for welding applications, the image processing time must be even faster.

High resolution systems are required in applications involving automated inspection and picture data retrieval where speed is sometimes not such an important criterion. This must not be confused with the needs of sensory robot systems.

The sensor crisis

Perhaps the key issue in the production of the sensory robot is the availability of suitable sensors. The following represents a summary of sensing requirements for robot applications:

 O presence,
 O range,
 O single axis measurement (or displacement),
 O two-dimensional location/position,
 O three-dimensional location/position,
 O thermal,
 O force.

for which the sensing devices or methods shown in Table 1 are available.

TABLE 1.
SENSING DEVICES AND METHODS AVAILABLE FOR ROBOTICS

Vision	*Tactile*
Photo-detector	Probe
Linear array	Strain gauge
Area array	Piezoelectric
TV camera	Carbon materials
Laser (triangulation)	Discrete arrays
Laser (time of flight)	Integrated arrays
Optical fibre	
Acoustic	*Other*
Ultrasonic detectors/emitters	Infra red
Ultrasonic arrays	Radar
Microphones (voice control)	Magnetic proximity
	Ionising radiation

The only satisfactory location for sensors is on the robot manipulator itself at or near the end effector.[14, 17] Locating an image sensor above the work area of a robot suffers from the disadvantage that the robot manipulator will obscure its own work area and the metric displacement of the end effector from its destination must be measured in an absolute rather than a relative way. Siting the sensor on the end effector allows relative measurements to be taken reducing considerably the need for calibration of mechanical position and the need for imaging linearity. Sensory feedback in this situation can be reduced to the simplicity of range finding in some applications.

What is missing from the sensor market are devices specifically tailored to be integrated close to the gripper jaws. The promise of solid-state arrays for this particular application has not materialised which is primarily due to the commercial incentives associated with the television industry. It might be accurate to predict that over the next decade imaging devices manufactured primarily for the television market will be both small and cheap enough to be useful in robot applications. However, at present, area array cameras are extremely expensive and, while smaller than most thermionic tube cameras, are far too large to be installed in the region of a gripper. Most of the early prototype arrays of modest resolution have been abandoned.

It is not an exaggeration to suggest that no imaging sensors exist which are ideally suited for robot applications. The use of dynamic RAM devices for image purposes[17] has proved to be a minor breakthrough and gives an indication of the rugged approach which is needed to achieve success. Some researchers have attacked the problem of size reduction by using coherent fibre optic to retrieve an image from the gripper area[16] which imposes a cost penalty on the total system. This approach can, however, exploit a fundamental property of optical fibre in that a bundle of coherent fibres can be sub-divided to allow a single high-resolution imaging device to be used to retrieve and combine a number of lower resolution images from various parts of the work area including the gripper – with each subdivided bundle associated with its own optical arrangements.[18, 19]

Linear arrays have been used in situations involving parts moving on a conveyor in such a way that mechanical motion is used to generate one axis of a two-dimensional image.[12, 18] There is no reason why the same technique should not be associated with a robot manipulated by using the motion of the end effector to generate a two-dimensional image. Also implied here is the possibility of using circular scanning of the work area or even taking a stationary image from a linear array.

Tactile sensing is required in situations involving placement. Both active and passive compliant sensors[20, 10] have not only been successfully demonstrated but have experienced a period of development and application in the field. The situation surrounding tactile array sensors is quite different. Because the tactile arrays are essentially discrete in design, they are inevitably clumsy and are associated with very low resolution. Interesting experiments have been reported[21-24] and an exciting development for a VLSI tactile sensor is to be published.[25]

Experiments with range sensing have been liberally researched and some success has been achieved with acoustic sensors and optical sensors (including lasers). The whisker probe[24] can now be replaced by a laser alternative with obvious advantages. Laser range finding is well developed but under-used in robot applications although the use of laser probes sited on the end effector of an industrial robot makes a natural automated dimensional inspection system.

It is clear from this brief catalogue of sensing methods that a great deal of chaos surrounds the sensor world. What we must work towards is some element of modularity in sensor design to allow for the optimum sensing method to be incorporated into a given application. No single sensor can provide the solution to all problems and bigger does not always mean better. However, the recent exciting developments in 'smart sensors' which incorporate primitive image processing (front-end processing) will be most welcome. A comprehensive survey and assessment of robotic sensors has been published by Nitzan.[26]

Languages and software
The involvement of the stored-program computer in the present generation of robots does no more than to provide an alternative to a hard-wired controller excepting that the computer provides an integrated memory facility to retain individual 'programs'.

Software and languages became a reality for most users with the introduction of VAL[27] which incorporates the capability to interpolate linear motion between two points and provides for co-ordinate transformation of axes. Further, VAL allows for transformation between vision and machine co-ordinates.[28] Machine training or learning using VAL can be achieved 'off-line' rather than the 'teach by showing' which is the method used by the majority of present generation robots.

Work is now under way on languages for assembly[29, 30] which will give to the sensory robot system autonomy of action within the requirements of the assembly task. Further, a common assembly language will permit the same instructions to be repeated on different robot hardware in the same way that computer programs written in a common language can be executed on different machines. The lesson to be learned here is that we must discipline ourselves to a common assembly language before a proliferation of language creates a situation disorder. It is still early days to consider an assembly task being executed by an 'off line' assembly language as part of a CAD/CAM operation. One of the stepping stones required in this revolutionary process is the establishment of some 'bench marks' to compare and test the relative merits of assembly languages.

Concluding remarks

The ingredients which comprise the second generation robot are:
- ○ mechanisms with speed and precision,
- ○ cheap and reliable sensors,
- ○ elegant and rugged software.

In all of these areas there exists a significant deficiency of development and perhaps a need for an innovative breakthrough. We have seen previously the short-fall in sensor requirements and the need for good supporting software has only been admitted in recent times. With the exception of mechanisms specifically designed for dedicated tasks, no existing robot device alone can really provide the precision necessary for assembly operations. A promising way to proceed is to use a programmable high speed manipulator for coarse positioning, coupled with a 'floating' table which incorporates features for fine positioning.[31]

When it is remembered that *research* demonstrations of 'generation 2' robots have been available over the past decade[32-37] it is salutory to recognise that predictions for the future made over this period have not been realised. A survey of the current situation in robot vision has been published recently.[38] It would be a brave person who now predicted where and when the 'generation 2' robot would take its place in industry. With the wisdom of hindsight we know that there is still a vast amount of research and development required coupled with an industrial need for a 'generation 2' robot. Perhaps this need will first appear in the textile, pottery or confectionery industries[39] to provide the 'shot in the arm' for 'generation 2' just as spot welding did for 'generation 1' a decade ago.

Over the past two years, the United Kingdom has introduced government funding to aid and support industrial applications as well as providing a co-ordination programme of research and development in the universities. Surveys of industrial requirements and research partnerships in the UK have recently been published.[40, 41]

References

[1] *Proc. 1st Int. Symp. on Industrial Robots,* Illinois Institute of Technology, Chicago, April 1970.

[2] J. F. Engelberger, *'Robotics in Practice'* Kogan Page, London, 1980.

[3] G. J. Gleason and G. J. Agin, 'A Modular Vision System for Sensor-controlled Manipulation and Inspection'. *Proc. 9th Int. Symp. on Industrial Robots*, SME, Washington, March 1979.

[4] P. F. Hewkin and M. A. D. Phil, 'OMS – Optical Measurement System' *Proc. 1st Int. Conf. Robot Vision and Sensory Controls,* IFS (Conferences) Ltd, Stratford-upon-Avon, England, April 1981.

[5] P. Villers, 'Present Industrial Use of Vision Sensors for Robot Guidance', *Proc. 12th Int. Symp. on Industrial Robots,* l'Association Française de Robotique Industrielle, Paris, June 1982.

[6] M. L. Baird, 'SIGHT-1': 'A Computer Vision System for Automated IC Chip Manufacture' *IEEE Trans. Systems, Man and Cybernetics,* Vol. 8, No. 2 (1978).

[7] S. Kawato and Y. Hirata, 'Automatic IC Wire Bonding System with TV Cameras', SME Technical Paper AD79-880, 1979.

[8] Johnston, E., 'Spray Painting Random Shapes Using CCTV Camera Control', *Proc. 1st Int. Conf. Robot Vision and Sensory Controls,* IFS (Conferences) Ltd, Stratford-upon-Avon, England, April 1981.

[9] H. Makino, N. Furuya, K. Soma and E. Chin, 'Research and Development of the SCARA robot', *Proc. 4th International Conference on Production Engineering,* Tokyo, 1980.

[10] D. E. Whitney and J. L. Nevins, 'What is the Remote Centre Compliance (RCC) and What Can it do?' *Proc. 9th Int. Symp. on Industrial Robots*, SME, Washington, March 1979.

[11] T. J. Fountain, V. Geotcherian, 'CLIP 4 Parallel Processing System', *IEEE Proc.*, Vol. 127, Pt. E, No. 5 (1980).

[12] S. W. Holland, L. Rossol and M. R. Ward, 'Consight-1: A Vision Controlled Robot System for Transferring Parts from Belt Conveyors'. *Computer Vision and Sensor Based Robots*, G. G. Dodd and L. Rossol, Eds. Plenum Press, New York, 1979.

[13] R. C. Bolles, 'Three Dimensional Locating of Industrial Parts', 8th NSF Grantees Conference on Production Research and Technology, Stanford, Calif., January 1981.

[14] R. N. Nagel, G. J. Vender-Brug, J. S. Albus and E. Lowenfeld, 'Experiments in Part acquisition Using Robot Vision', SME Technical Paper MS79-784 1979.

[15] P. Saraga and B. M. Jones, 'Parallel Projection Optics in Simple Assembly', *Proc. 1st Int. Conf. Robot Vision and Sensory Controls'*, IFS (Conferences) Ltd, Stratford-upon-Avon, England, 1981.

[16] I. Mazaki, R. R. Gorman, B. H. Shulman, M. J. Dunne and H. Toda, 'Arc Welding Robot with Vision', *Proc. 11th Int. Symp. on Industrial Robots*, JIRA, Tokyo, October 1981.

[17] P. M. Taylor, G. E. Taylor, D. R. Kemp, J. Stein and A. Pugh, 'Sensory Gripping System: The Software and Hardware Aspects' *Sensor Review*, Vol. 1, No. 4 (October 1981).

[18] A. J. Cronshaw, W. B. Heginbotham and A. Pugh, 'Software Techniques for an Optically-tooled Bowl Feeder', *IEEE Conference 'Trends in On-line Computer Control Systems'*, Sheffield, England, Vol. 172, March 1979.

[19] J. H. Streeter, 'Viewpoint – Vision for Programmed Automatic Assembly', *Sensor Review*, Vol. 1, No. 3 (July 1981).

[20] T. Goto, et. al., 'Precise Insert Operation by Tactile Controlled Robot', *The Industrial Robot*, Vol. 1, No. 5 (Sept. 1974).

[21] M. H. E. Larcombe, 'Carbon Fibre Tactile Sensors', *Proc. 1st Int. Conf. Robot Vision and Sensory Controls*, IFS (Conferences) Ltd., Stratford-upon-Avon, England, April 1981.

[22] J. A. Purbrick, 'A Force Transducer Employing Conductive Silicone Rubber' Ibid.

[23] N. Sato, W. B. Heginbotham and A. Pugh, 'A method for three-dimensional Robot Identification by Tactile Transducer', *Proc. 7th Int. Symp. on Industrial Robots*, JIRA, Tokyo, October 1977.

[24] S. S. M. Wang and P. M. Will, 'Sensors for Computer Controlled Mechanical Assembly', *The Industrial Robot*, Vol. 5, No. 1 (March 1978).

[25] J. E. Tanner and M. H. Raibert, 'A VLSI Tactile Array Sensor', *Proc. 12th Int. Symp. on Industrial Robots*, l'Association Française De Robotique Industrielle, Paris, June 1982.

[26] D. Nitzan, 'Assessment of Robotic Sensors', *Proc. 1st Int. Conf. Robot Vision and Sensory Controls*, IFS (Conferences) Ltd, Stratford-upon-Avon, England, April 1981.

[27] 'Users Guide to VAL – Robot Programming and Control System', Unimation Inc., Danbury, Conn.

[28] B. Carlisle, 'The PUMA/VS-100 Robot Vision System', *Proc. 1st Int. Conf. Robot Vision and Sensory Controls*, IFS (Conferences) Ltd., Stratford-upon-Avon, England, April 1981.

[29] L. I. Leiberman and M. A. Nelsey, 'AUTOMPASS – An Automatic Programming System for Computer Controlled Mechanical Assembly', *IBM Journal of Research and Development*, Vol. 21, No. 4 (1977).

[30] R. J. Popplestone, et. al., 'RAPT: A Language for Describing Assemblies', University of Edinburgh, UK, September 1978.

[31] J. Hollingum, 'Robotics Institute Teams Development in University and Industry'. *The Industrial Robot*, Vol. 8, No. 4, December 1981.

[32] W. B. Heginbotham, D. W. Gatehouse, A. Pugh, P. W. Kitchin and C. J. Page, 'The Nottingham SIRCH Assembly Robot', *Proc. 1st Conf. on Industrial Robot Technology*, IFS Ltd., Nottingham, England, March 1973.

[33] Y. Tosuboi and T. Inoue, 'Robot Assembly Using TV Camera', *Proc. 6th Int. Symp. on Industrial Robots'*. IFS Ltd., Nottingham, England, March 1976.

[34] D. Nitzan, 'Robotic Automation at SRI' *Proc. of MIDCON/79*, Chicago, Illinois, November 1979.

[35] P. Saraga and D. R. Skoyles, 'An Experimental Visually Controlled Pick and Place Machine for Industry', *Proc. 3rd Int. Conf. on Pattern Recognition*, IEEE Computer Society, Coronado, Calif., November 1976.

[36] D. A. Zambuto and J. E. Chaney, 'An Industrial Robot with Mini-Computer Control', *Proc. 6th Int. Symp. on Industrial Robots*, IFS (Conferences) Ltd., Nottingham, England, March 1976.

[37] J. R. Birk, R. B. Kelley and H. A. S. Martins, 'An Orientating Robot for Feeding Workpieces Stored in Bins', *IEEE Trans. Systems Man and Cybernetics*, Vol. 11, No. 2 (1981).

[38] R. P. Kruger and W. B. Thompson, 'A Technical and Economic Assessment of Computer Vision for Industrial Inspection and Robotic Assembly' *Proc. IEEE*, Vol. 69, No. 12 (1981).

[39] A. J. Cronshaw, 'Automatic Chocolate Decoration by Robot Vision', *Proc. 12th Int. Symp. on Industrial Robots*, l'Association Française de Robotique Industrielle, Paris, June 1982.

[40] J. King and E. Lau, 'Robotics in the UK', *The Industrial Robot*, Vol. 8, No. 1 (March 1981).

[41] P. G. Davey, 'UK Research and Development in Industrial Robots' *2nd International Conference on 'Manufacturing Matters'*, The Institution of Production Engineers, London, England, March 1982.

COMPUTER VISION IN INDUSTRY

Lothar Rossol, General Motors Research Laboratories, USA

Presented at the Meadowbrook Research Conference 'Robotics, The Next Decade', May 25, 1982, Oakland University, Rochester, Michigan, USA

Introduction

Computer vision will revolutionise manufacturing in two important areas: inspection and adaptive control of robots. Already both these uses of computer vision are beginning to change manufacturing, with many systems now in production. The two specific industrial vision systems to be described in this paper, KEYSIGHT and CONSIGHT, will give a general indication of the state of the technology.

Robotics and computer vision are classical examples of rapid change in a high-technology field, with advanced systems continually obsoleting existing products. Further major advances can safely be anticipated for the next decade and beyond. Industry has led in the development of new vision and robot technology and is in a position to make many of the technological advances in the future as well. This paper will attempt to forecast some of these advances.

Computer vision

The typical computer vision system includes an electronic camera, a computer, and software. Much can be done with such systems; however, we cannot yet claim that computers or robots can actually 'see', at least not in the usual meaning of the word. Because neither biological nor artificial vision is well understood, or easy to describe, we tend to characterise vision machines in terms of peripheral issues, such as speed, accuracy, and cost. These issues are important, but from a research viewpoint they are not central. The heart of the problem is neither the camera, nor the architecture of the computer, but the software. Most industrial vision tasks are not solvable today because we simply do not know how to write the programs. (More precisely stated, we lack appropriate algorithms, since software can always be implemented in special purpose electronics.)

We can characterise the universe of vision machines as shown in Fig. 1.

Essentially all vision machines operate in the $y = 0$ plane, that is, they are restricted to shades of grey and cannot deal with colour picture data. Furthermore, commercially available machines are positioned along the z-axis, that is, they are capable of processing black and white or binary picture data only. Tasks of substantial complexity can be solved with these machines, so long as the visual scene can be thresholded, resulting in a binary picture.

The difficulty is, of course, that the majority of industrial tasks do not fall on the z-axis. Significant variations in light reflectance from part to part will usually

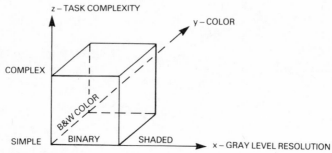

Fig. 1. Dimensions of computer vision.

complicate the task. The area of interest is often neither predictably brighter nor darker than the surrounding area. Furthermore, it is seldom possible or practical to design appropriate lighting to make scenes of many colours and many different shades of grey appear of consistent high contrast black and white to the camera. The major problem in industrial vision tasks, that of object-background separation, or segmentation, is thus generally not solvable by simple thresholding.

GM, Hitachi, and others have built vision systems capable of processing grey level picture data, that is, systems positioned farther along the x-axis of Fig. 1. KEYSIGHT, described in this paper, is an example of such a system. However, these systems operate close to the z = 0 plane, that is, task complexity is low. Unfortunately, however, the majority of industrial tasks are of high complexity.

We are thus left with a dilemma. We can build sophisticated binary vision systems for a limited market. Or we can address the substantial market for grey level vision systems. But at this time we still lack much of the necessary technology that will allow us to build grey level systems of sufficient power and generality.

Research will of course continue to extend the state of the technology in grey level vision. There is, however, an alternate approach worthy of consideration. This approach promises not only to solve the ubiquitous problem of segmentation in vision, but it will also allow part geometry to be sensed directly, as opposed to inferring it as in grey level picture processing. It will further more substantially reduce the amount of data that has to be processed, a difficult and costly engineering issue in grey level systems.

The central idea in this alternate approach is the use of structured light. Here a carefully designed light pattern is projected into the scene, and then viewed with the camera from a different direction. Such a system differs from conventional binary systems in that distortions in the pattern are used to calculate directly the geometric shape of the objects in the scene. Although intellectually perhaps not as appealing as other methods because of dissimilarities to biological systems, structured light systems nevertheless will be of substantial use to industry.

This paper will briefly describe two state of the art vision systems, CONSIGHT and KEYSIGHT. CONSIGHT uses structured light, while KEYSIGHT processes grey level picture data.

Consight

Consight[1] is a visually-guided robot system. It transfers parts from a moving conveyor belt. Part transfer operates as follows: The conveyor carries the randomly positioned parts past a vision station which determines each part's position and orientation on the moving belt. This information is sent to a robot system which tracks the moving parts, picks them up, and transfers them to a predetermined location.

CONSIGHT can handle a continuous stream of parts on the belt, so long as these parts are not touching. Maximum speed limitation is imposed by the cycle time of the robot arm, rather than by the vision system or the computer control.

The vision system employs a linear array camera that images a narrow strip across the belt perpendicular to the belt's direction of motion. Since the belt is moving it is possible to build a conventional two-dimensional image of passing parts by collecting a sequence of these image strips. Uniform spacing is achieved between successive strips by use of the belt position detector which signals the computer at appropriate intervals to record the camera scans of the belt.

A narrow and intense line of light is projected across the belt surface. The line camera is positioned so as to image the target line across the belt. When an object passes into the beam, the light is intercepted before it reaches the belt surface [Fig. 2]. When viewed from above, the line appears deflected from its target wherever a part is passing on the belt. Therefore, wherever the camera sees brightness, it is viewing the unobstructed belt surface; wherever the camera sees darkness, it is viewing the passing part.

Unfortunately, a shadowing effect causes the object to block the light before it actually reaches the imaged line, thus distorting the part image. The solution is to use two light sources directed at the same strip across the belt. When the first light source is prematurely interrupted, the second will normally not be. By using multiple light sources and by adjusting the angle of incidence appropriately, the problem is essentially eliminated.

Since several pieces of an object or even different objects may be underneath the camera at any one time, the continuity from line to line of input must be monitored. CONSIGHT uses 6-connected regions to establish continuity. That is, connectivity is permitted along the four sides of a picture element and along one of the diagonals.

Once the passing objects have been isolated from the background, they may be analysed to determine their position and orientation relative to a reference coordinate system.

For each object detected, a small number of numerical descriptors is extracted. Some of these descriptors are used for part classification, others for position determination.

The part's position is described by the triple (x, y, theta). The x and y values are always selected as the centre of area of the part silhouette. There is no convenient method for uniquely assigning a theta value (rotation) to all parts. A useful descriptor for many parts, however, is the axis of the least moment of inertia (of the part silhouette). For long thin parts this can be calculated accurately. The axis must still be given a sense (i.e., a direction) to make it unique, a process that can be accomplished in a variety of ways, but is part-specific.

The computer model for the part specifies the manner in which the theta value should be computed. For example, one method available for giving a sense to the axis value is to select the moment axis direction that points nearest to the maximum radius point measured from the centroid to the boundary. Another method uses the centre of the largest internal feature (e.g., a hole) to give direction to the axis. A more general technique based on third order moments is also available.

Keysight

KEYSIGHT is an example of a computer vision system that processes grey level picture data.[2] It inspects valve spring assemblies on engine heads for the presence

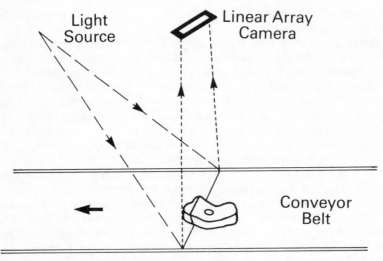

Fig. 2. Camera is positioned to image target line across the belt.

of valve spring cap keys. A V8 engine head (two per engine) has eight valve spring assemblies (four intake and four exhaust) as shown in Fig. 3. Fig. 4a shows a good assembly with both keys present while Fig. 4b shows an assembly with one missing key. The KEYSIGHT system inspects each assembly for the presence of both keys. Since the variation in assembly location is greater than the half-width of the keys, KEYSIGHT must first find the exact location of the assembly before it can inspect for missing keys.

Significant variations in the reflectances of different valve spring assemblies further complicate the task. The keys have different colours to indicate different hardnesses; the stems are sometimes painted with various colours; an oil film may cover the keys; because of an extra indentation, the intake and exhaust valves have a different appearance. Furthermore, reliability of the system over thousands of operations is crucial. This reliability requirement, typical of production systems, has considerable impact on the design of vision algorithms. It is not sufficient that these systems work correctly only most of the time, instead the algorithms must be such that failure is essentially impossible.

KEYSIGHT digitises a picture of spatial resolution 64 × 64 points with a grey level resolution of four bits. The field of view of the camera system is slightly larger than the assembly.

To find the centre of the assembly, KEYSIGHT first calculates the gradient at every picture point, using a simple absolute-value approximation to the gradient operator. It then applies the thinning operation described next to the gradient picture.

Edge pixels are separated into either horizontal or vertical edges, depending on which component of the edge is larger. If an edge pixel is horizontal, its two neighbours (above and below) are compared with it. If the edge magnitude at the point is smaller than that of either neighbour, the edge magnitude is set to zero. For vertical pixels a similar adjustment is made. The results are stored in a new array, resulting in a thinning operation that is not sequence dependent. Next the thinned edge pixels are compared to a fixed threshold, and those edge pixels exceeding the threshold are used to locate the centre of the assembly. The centre of the assembly corresponds to the centre of symmetry of the thresholded and thinned edge pixel distribution. A two-dimensional convolution function locates the centre of symmetry and thus the centre of the assembly.

Since the convolution method will result in a centre location for any arbitrary input picture, it is necessary to verify that the centre of an actual assembly has been found. Because the centre location is known, a mask (or template) can conveniently be used for verification. Use of a grey level mask, however, is inappropriate here because the relative brightness of the stem and keys is not known a priori. Consequently an edge mask is used and compared to the image edge pattern. The 'goodness' of the fit is simply the total number of mask and image edge points that have the same location.

Once the centre of the assembly has been found, key inspection can take place. The keys are consistently brighter than the hole that results when a key is missing. The program checks for the keys by examining the intensity profile along the circumference of a circle through the keys. A threshold is calculated using the average intensity of points along the circle. A valve spring assembly passes inspection if the percentage of the circle that is dark lies in a specified range. This simple test works consistently well in spite of the fact that there are variations in contrast between key and hole, and that occasionally special marks on the keys will cause black spikes in the intensity profile.

Several KEYSIGHT systems are now used in production. They typify the sophistication possible for grey level vision systems.

Discussion

KEYSIGHT and CONSIGHT are representative of vision systems now used in production. KEYSIGHT is typical of vision for inspection; CONSIGHT of vision for robot guidance. KEYSIGHT demonstrates the sophistication now achievable with grey level picture processing; CONSIGHT demonstrates the use of structured light.

The structured lighting of CONSIGHT converts a difficult grey level vision problem into an inherently simpler binary problem. CONSIGHT, however, differs from conventional binary systems in that 3-dimensional structure can be calculated directly, rather than having to be inferred from light reflectivities.

CONSIGHT uses structured light in a highly restricted way, since it resolves depth or range into a 2-valued or 'binary' function only. An area camera used with CONSIGHT lighting could resolve depth into arbitrarily many values, limited only by the resolution of the camera. An additional extension, structured light used in conjunction with stereo vision, should result in highly reliable 3-dimensional vision systems.

KEYSIGHT demonstrates a shortcoming of grey level inspection systems: they are not easily programmable, especially when compared to binary systems. KEYSIGHT can inspect for valve keys – nothing else. A major research objective is to build 'teachable' grey level vision systems.

Binary vision, the remaining major category, is now becoming available in commercial machines. Furthermore, binary vision has also been developed for internal use by many companies, including GM, Hitachi, and Texas Instruments, and hence this category requires little further discussion.

Expected advances

In the future we can expect to see major advances in structured light and grey level vision. These advances will be driven by the need for inspection and for sensor-controlled robots in industry. In particular, the desire to eliminate precise positioning requirements is rapidly creating a market for robots with vision, force, and range-finding capabilities. Of these three sensors, vision is likely to have the most profound effect.

We can draw the following conclusions regarding vision (and sensors in general): Sensory input to robots (vision, force, range) will eliminate precise positioning requirements and provide real-time workplace adaptability and general inspection capability. Vision systems will provide positioning information to robots so that parts, robots, and surrounding equipment need not be precisely located. Vision systems will perform a range of inspection tasks including incoming part inspection, in-process inspection, and final quality control inspection. In a related area, force sensing and control systems will provide fine positioning control, allow for the detection of unusual conditions such as collisions, and make possible high-precision assembly.

The table below summarises the effects of the above advances in terms of new functional capabilities. Included are rough estimates on when these functions might become generally available. We do not expect all of these features to be available in the predicted times, however, any of them could be available given appropriate demand and R&D resources.

Expected sensor capabilities

	General Availability
ROBOT PART FEEDING	
Floor Conveyors	1984
Pallets and Trays	1986
Overhead Conveyors	1987
Bins	1988
INSPECTION	
Simple Inspection	1983
Programmable Inspection	1986
Complex Inspection	1988
ADAPTIVE CONTROL FOR ROBOTS	
Line Tracking	1983
Automatic Seam Tracking	1984
Simple Force	1984
In-process Monitoring	1985
Full Force	1987

Robot Part Feeding
Part feeding, without the use of specially designed equipment for precisely locating parts, will become an integral part of the robot system. Vision and range sensors

Fig. 3. Engine head.

Fig. 4. Valve spring assemblies. (a) Both keys present.   (b) One key missing.

will provide the means for a robot to acquire randomly located parts from belt conveyors, off pallets, out of bins, and off overhead conveyors.

Inspection

Computer vision systems will be able to perform a variety of inspection tasks. Simple inspection tasks, such as checking for part presence, are being done now with computer vision, but sophisticated grey level inspection systems are not generally available from vendors. By programmable inspection systems is meant grey level or structured light systems that can be easily reprogrammed for a variety of different parts, similar to binary systems available now. By complex inspection is meant a qualitative inspection, that is, a test for the 'goodness' of a component or assembly. Examples of complex inspection tasks are: PC board inspection, metal finish inspection, and sub-assembly inspection.

Adaptive Control for Robots

Adaptive control is defined as the use of sensory data for real-time control of the robot operations. In-process monitoring involves monitoring an on-going process and adapting the robot operation (perhaps speed). Automatic seam tracking would be used in arc welding and adhesive application. Simple force sensing utilises a 1 or 2-axis sensor to measure forces (torques) along 1 or 2 dimensions. Complex force sensing measures and controls all six forces and torques.

Conclusions

The field of computer vision has been a prime example of high technology in that advanced vision systems have continually obsoleted existing products. Further major advances in computer vision, particularly in structured light and grey level vision, are predicted for the next decade and beyond. These advances will revolutionise manufacturing approaches to inspection and robotics.

Acknowledgements

CONSIGHT is the result of the dedicated efforts of many individuals, including Mitch Ward and Steve Holland. The ideas in KEYSIGHT are largely due to Walt Perkins. Mitch Ward contributed substantially to predictions on advances in computer vision and robotics. The idea of the 'vision cube' is due to Steve Holland.

References

[1] M. R. Ward, L. Rossol, S. W. Holland, 'CONSIGHT: An Adaptive Robot with Vision', *Robotics Today* pp. 26–32 (Summer 1979).

[2] L. Rossol, 'Vision and Adaptive Robots in General Motors', *Proc. 1st Int. Conf. on Robot Vision and Sensory Controls*, Stratford-upon-Avon, UK, IFS (Publications) (April 1981).

[3] C. A. Rosen, 'Machine Vision and Robotics: Industrial Requirements', *Computer Vision and Sensor-Based Robots*, ed. G. G. Dodd and L. Rossol, Plenum Press (1979).

Chapter 2
VISUAL PROCESSING TECHNIQUES

Visual sensors can produce vast quantities of data which needs efficient processing methods if their processing time is to be compatible with industrial machine cycle times. The four papers outline in detail some of the techniques available and how they may be evaluated.

PROCESSING OF BINARY IMAGES

P. W. Kitchin, Patscentre Benelux, Belgium and A. Pugh, Department of Electronic Engineering, University of Hull, England

First presented at 1st SERC Vacation School on Robotics, University of Hull, September, 1981. Reproduced by kind permission of Science and Engineering Research Council.

Summary

This paper is tutorial in nature and gives a treatment of the implementation of edge following algorithms based on Freeman chain coding.[4-9] Binary image processing has direct relevance to applications of robot vision even if the original picture is captured as a grey scale image. It is particularly difficult to identify publications such as this which give a detailed account of the methods used in the technology of robot vision. These notes are by no means comprehensive and represent part of the research studies of a single group over the period 1969-73. The unconventional nomenclature retained in this paper is identical to that used in the related Ph.D thesis published in 1977.[22] This facilitates reference to this more extensive treatment of the subject.

Introduction

A considerable amount of research has been reported into methods of processing two dimensional images stored as binary matrices (for bibliographies see Ullman,[1] Rosenfeld[2]). A large part of this work has been directed towards solving problems of character recognition. Whilst many of these techniques are potentially useful in the present context, it is valuable to note some important differences between the requirements of character recognition and those associated with visual feedback for mechanical assembly.

□ *Shape and size.* All objects presented to the assembly machine are assumed to be exact templates of the reference object. The objects may be any arbitrary geometric shape, and the number of possible different objects is essentially unlimited. Any deviation in shape or size, allowing for errors introduced by the visual input system, is a ground for rejection of the object (though this does not imply the intention to perform 100% inspection of components). The derived description must therefore contain all the shape and size information originally present in the stored image. A character recognition system has in general to tolerate considerable distortion ('style') in the characters to be recognised, the most extreme example being the case of handwritten characters. The basic set of characters is, however, limited. The closest approach to a template matching situation is achieved with the use of type fonts specially designed for optical reading, such as OCR A and OCR B.

☐ *Position and orientation.* A component may be presented to the assembly machine in any orientation and any position in the field of view. Though a position and orientation invariant description is required in order to recognise the component, the measurement of these parameters is also an important function of the visual system to enable subsequent manipulation. While lines of characters may be skewed or bowed, individual characters are normally presented to the recognition system in a relatively constrained orientation, a measurement of which is not required.

☐ *Multiple objects.* It is a natural requirement that the visual system for an assembly machine should be able to accommodate a number of components randomly positioned in the field of view. The corresponding problem of segmentation in character recognition is eased (for printed characters) by a priori knowledge of character size and pitch. Such information has enabled techniques for the segmentation of touching characters.[1] No attempt is made in this study to distinguish between touching objects. Their combined image will be treated by the identification procedures as that of a single, supposedly unknown object.

The essentially unlimited size of the set of objects that must be accommodated by the recognition system demands that a detailed description of shape be extracted for each image. There are, however, a number of basic parameters which may be derived from an arbitrary shape to provide valuable classification and position information.

These include:

○   area,

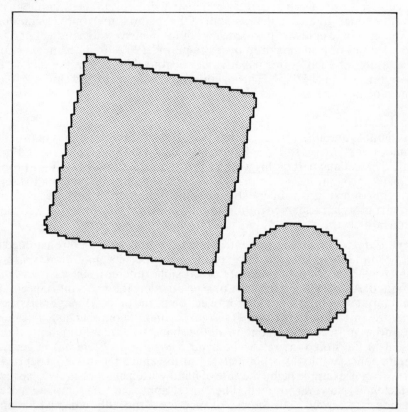

Fig. 1. The frame after procedure INPUT-FRAME.

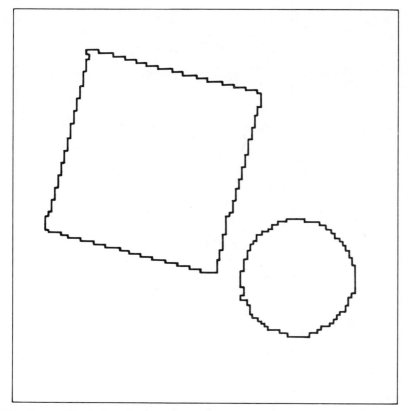

Fig. 2. The frame of Fig. 1 after edge extraction.

○ perimeter,
○ minimum enclosing rectangle,
○ centre of area,
○ minimum radius vector (length and direction),
○ maximum radius vector (length and direction),
○ holes (number, size, position).

Measurements of area and perimeter provide simple classification criteria which are both position and orientation invariant. The dimensionless 'shape' factor (area)/(perimeter)[2] has been used as a parameter in objection recognition.[3] The coordinates of the minimum enclosing rectangle provide some information about the size and shape of the object, but this information is orientation dependent. The centre of area is a point that may be readily determined for any object, independent of orientation, and is thus of considerable importance for recognition and location purposes. It provides the origin for the radius vector, defined here as a line from the centre of area to a point on the edge of an object. The radius vectors of maximum and minimum length are potentially useful parameters for determining both identification and orientation. Holes are common features of engineering components, and the number (if any) present in a part is a further suitable parameter.

The holes themselves may also be treated as objects, having shape, size and position relative to the object in which they are found.

The requirements for the establishment of connectivity in the image and the

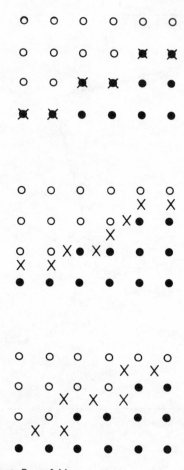

a. The border according to Rosenfeld.
b. The edge according to Rosenfeld.
c. The scheme adopted.
Fig. 3. Some possible outline definitions.

derivation of detailed descriptions of arbitrary geometric shapes are most appropriately met by an edge-following technique. The technique starts with the location of an arbitrary point on the black/white edge of an object in the image (usually by a raster scan). An algorithm is then applied which locates successive connected points on the edge until the complete circumference has been traced and the starting point is reached. If the direction of each edge point traced relative to the previous point is recorded, a one-dimensional description of the object is built up which contains all the information present in the original shape. Such chains of directions have been extensively studied by Freeman.[4-9] Measurements of area, perimeter, centre of area and enclosing rectangle may be produced while the edge is being traced, and the resulting edge description is in a form convenient for the calculation of radius vectors.

Edge-following establishes connectivity for the object being traced. Continuing the raster scan in search of further objects in the stored image then presents the problem of the 're-discovery' of the already traced edge.

Rosenfeld[2] suggests changing all edge points visited by the edge-following routine from '1' to '2's for the first object traced, to '3's for the second object, and so on (where a '1' represents an object, or 'black' in the original image, and '0' represents the background or 'white'). This approach, however, necessitates the provision of more than one bit of storage for each picture point in the image and is consequently not appropriate to a system in which minimising memory usage is important. The solution adopted by the author is to apply a spatially differentiating or edge extracting operator to the image immediately after input from the camera. This operator replaced all '1's in the image by '0's unless they are deemed to lie on a black/white border. A computer plot of the contents of FRAME with the camera viewing a square and a disc is shown in Fig. 1, and the result of applying the edge-extracting operator is shown in Fig. 2. The edge-following procedure, called EDGETRACE, may now be applied to the image in the same way as for 'solid' objects. The procedure is arranged, however, to reset each edge point as it is traced. The tracing of a complete object thus removes it from the image, and ensures that it will not be subsequently retraced.

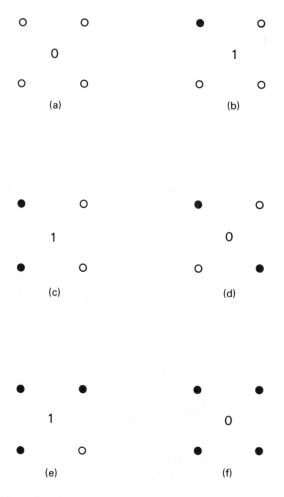

Fig. 4. Definition of the outline elements.

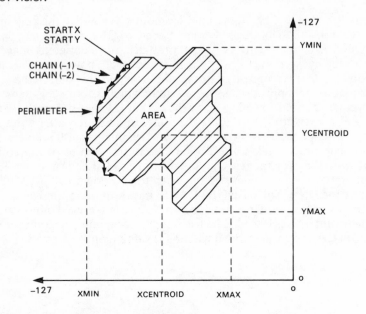

Fig. 5. Parameters returned by procedure EDGETRACE.

Smoothing

Picture processing techniques based on edge-following are likely to fail if the image contrast boundaries are broken. Such operations in character recognition systems are consequently often preceded by a smoothing or local averaging process, which tends to eliminate voids (white elements that should be black) and spurious black points, and to bridge small gaps. In the smoothing technique due to Dineen[10] an n × n element window is positioned over each element in the image in turn. The total number of black elements in the window are counted. A new image is formed in which each element corresponds to one position of the window, and each element is black only if the number of black elements in the corresponding window exceeds a preset threshold. The smallest practicable window size is 3 × 3 elements. Unger[11] has proposed a smooth process in which, instead of averaging, logical rules are applied to the contents of a 3 × 3 window, in order to determine the value of the corresponding element in a new pattern. The disadvantage of both systems in the present context is the relatively large amount of computer bit manipulation involved. Each requires of the order of 100 computer instructions to create a new pattern element.

The objects viewed by the assembly machine, however, are fundamentally different from printed characters, in that any gaps or voids in the objects themselves represent defects, and would be valid grounds for non-recognition. Isolated noise points occur infrequently, and will not in general affect the edge-following algorithm. Consequently, though steps are taken in the following procedures to minimise the effect of noise points, no specific smoothing operation is performed.

Edge extraction – procedure OUTLINEFRAME

The edge extraction operator must produce an unbroken sequence of edge points to ensure the success of the subsequent edge-following program. To minimise computing time, the generation of each element for the new FRAME must involve accessing the smallest possible number of picture elements in the original FRAME.

Rosenfeld[12] distinguishes between the 'border' and the 'edge' of a pattern as shown in Fig. 3, defining the border as being made up of the outermost elements of the pattern, and the edge as lying midway between horizontally or vertically adjacent pairs of pattern/background points. The concept of 'edge' rather than 'border' has been adopted for this application as it more accurately represents the true boundary between black and white in the physical world.

It also provides an unambiguous solution to the case of a line of single point thickness. The border concept would produce a result unacceptable to a subsequent edge-following routine, as the two boundaries of the line would be coincident.

A disadvantage of the edge as defined by Rosenfeld is that the resulting array of points has twice the point density of the original array, with on one-to-one relationship between the two arrays. A new array of edge points is therefore defined, having the spacing and density of the original, but with each point shifted in both X and Y directions by one half-point spacing. Each point in the new array lies at the centre of a square formed by four points in the original array, and its state (black or white – edge or not edge) is derived from the states of these four points. A similar solution has been discussed by Saraga and Wavish.[13]

The six basic configurations of the sixteen possible sets of four binary picture points are shown in Fig. 4, together with the intuitively determined values of the associated points in the edge array. Configuration 'd' is chosen not to yield an edge point in order to ensure the unambiguous treatment of single point thickness lines.

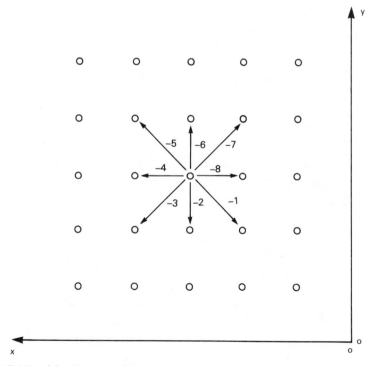

Fig. 6. Definition of the elemented chain vectors.

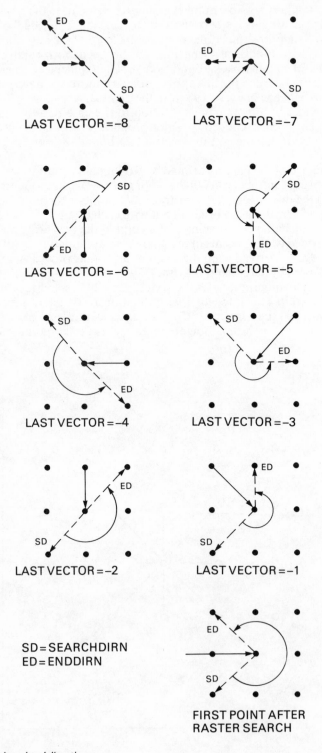

LAST VECTOR = −8

LAST VECTOR = −7

LAST VECTOR = −6

LAST VECTOR = −5

LAST VECTOR = −4

LAST VECTOR = −3

LAST VECTOR = −2

LAST VECTOR = −1

SD = SEARCHDIRN
ED = ENDDIRN

FIRST POINT AFTER
RASTER SEARCH

Fig. 7. New search and end directions.

Edge following – procedure EDGETRACE
Parameters
EDGETRACE has been written to provide the basic function of determining whether a point located in the differentiated FRAME forms part of the outline of an object. The coordinates of the initial point are loaded into the global integers START X, START Y. EDGETRACE is then called as a conditional procedure, that is in a statement of the form.

IF EDGETRACE THEN . . . . ELSE

The 'condition true' return from the procedure occurs if a closed outline greater than a preset minimum length is successfully traced.

EDGETRACE provides a number of parameters concerning the outline, together with a description of the outline in the form of a list of vectors, stored as global variables, (Table 1). These parameters are illustrated in Fig. 5.

The generation of the values XMAX, XMIN, YMAX, YMIN has been made optional. If they are required, the flag word EDGEFLAG is set (that is, made non-zero). Similarly the recording of CHAINCOUNT and the array CHAIN may be allowed or suppressed using the word CHAINFLAG.

Chain notation and the basic algorithm
For the purposes of this study, a point in the image matrix is defined as being connected to another point if it occupies one of the eight immediately adjacent locations in the matrix. The path between any pair of connected points may be denoted by one of eight elemental vectors (Fig. 6) whose directions have been labelled –8 to –1 (negative to facilitate indexing in PL-516).

An edge-following algorithm causes each point found in the outline to be reset to '0' before the search moves on to the next point. An outline is thus eliminated from the stored matrix FRAME as it is traced. The base for the search progresses around an outline in an anti-clockwise direction, provided that the outline is first located from outside its enclosed area by the raster-type search. The outline can only be

TABLE 1. PARAMETERS RETURNED BY EDGETRACE

VARIABLE	*VALUE*
PERIMETER	The length of the traced outline (a positive integer)
AREA	The enclosed area (a positive integer)
XMAX XMIN YMAX YMIN	The maximum and minimum X and Y coordinates reached by the outline (negative integers)
XCENTROID YCENTROID	The X and Y coordinates of the centre of the enclosed area (negative integers)
CHAINCOUNT	The number of elemental vectors making up the traced outline; also equal to the number of outline points traced (a negative integer)
CHAIN	An ordered array of the directions on the vectors making up the outline. The array has CHAINCOUNT elements.

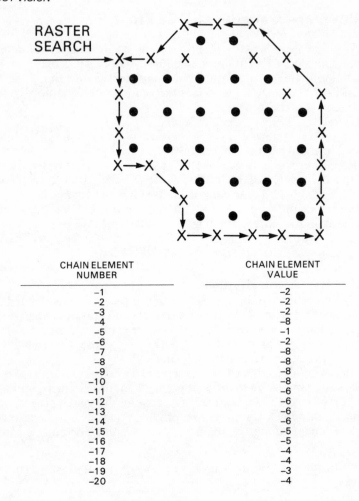

CHAIN ELEMENT NUMBER	CHAIN ELEMENT VALUE
−1	−2
−2	−2
−3	−2
−4	−8
−5	−1
−6	−2
−7	−8
−8	−8
−9	−8
−10	−8
−11	−6
−12	−6
−13	−6
−14	−6
−15	−5
−16	−5
−17	−4
−18	−4
−19	−3
−20	−4

Fig. 8. The chain description of a simple outline.

located from inside if it is intersected by the edge of the image matrix. In this case, the procedure EDGETRACE will execute 'condition false' return.

From each base position not all the eight adjacent points need be tested, as some will have been tested from the previous base position. Before starting a local search, the first direction to be tested (SEARCHDIRN) is preset to value depending on the previous vector. The final direction to be tested before the search is abandoned (ENDDIRN) is also preset.

The preset directions are shown in Fig. 7, together with the values for the case of the first search after entering the routine, when the previous search would have been in a raster pattern. It can be seen that up to six points have to be tested in each search. However, the number of tests will normally be smaller. In the case of the straight section of outline, only one or two points will be tested from each base position before the next outline point is found.

A complete description of the outline is formed by listing the directions of the vectors linking one point to the next, in the order that they are determined (the array 'CHAIN'). The result of applying the edge-following procedure to a simple outline is shown in Fig. 8, together with the derived 'chain' description.

As each element in an outline is located calculations leading to the determination of perimeter, area and first moments of area are updated.

Calculation of perimeter

The perimeter is measured as the sum of the magnitudes of the constituent elemental vectors. The vectors have magnitudes of 1 unit (even-numbered directions) or $\sqrt{2}$ units (odd-numbered directions).

The image matrix is assumed to be of unit spacing (Fig. 6). As each point in the outline is located, one of the two integers EVENPERIM or ODDPERIM is incremented, depending on the last vector direction. At the end of the procedure the calcuation:

$$\text{PERIMETER} = \text{EVENPERIM} + \text{ODDPERIM}.\sqrt{2}$$

is performed, the result being rounded to the nearest whole number.

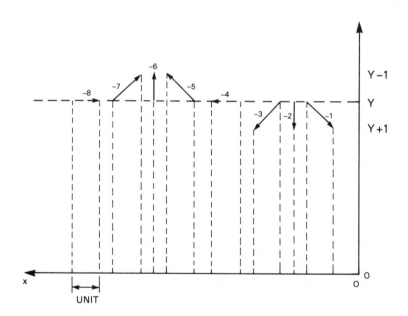

VECTOR DIRECTION	$\triangle$ AREA	$2.\triangle$ AREA
–8	$Y.1$	$2Y$
–7	$(Y-\frac{1}{2}).1$	$2Y-1$
–6	0	0
–5	$(Y-\frac{1}{2})(-1)$	$-2Y+1$
–4	$Y.(-1)$	$-2Y$
–3	$(Y+\frac{1}{2})(-1)$	$-2Y-1$
–2	0	0
–1	$(Y+\frac{1}{2}).1$	$2Y+1$

Fig. 9. Element contribution for area calculation.

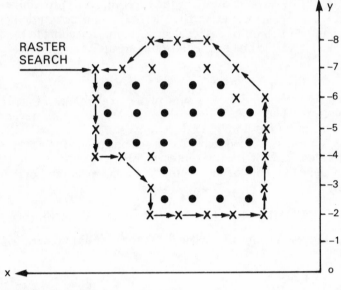

CHAIN ELEMENT NUMBER	CHAIN ELEMENT VALUE	2△ AREA	Σ2△ AREA
−1	−2	0	0
−2	−2	0	0
−3	−2	0	0
−4	−8	−8	−8
−5	−1	−7	−15
−6	−2	0	−15
−7	−8	−4	−19
−8	−8	−4	−23
−10	−8	−4	−31
−11	−6	0	−31
−12	−6	0	−31
−13	−6	0	−31
−14	−6	0	−31
−15	−5	+13	−18
−16	−5	+15	−3
−17	−4	+16	+13
−18	−4	+16	+29
−19	−3	+15	+44
−20	−4	+14	+58

GIVING AREA = 29

Fig. 10. The area calculation for a simple outline.

Calculation of area

The area enclosed by an outline is measured by summing the areas between each elemental vector and a convenient line (chosen to be the line $Y = O$). The sign convention adopted results in vectors having a decreasing x component (directions −5, −4, −3) contributing positive areas, and vectors having an increasing x component (directions −7, −8, −1) contributing negative areas (Fig. 9 – note that Y is a negative number). The net area enclosed by an outline traced in an anti-clockwise direction is then positive.

A simple example is shown in Fig. 10. To avoid unnecessary manipulation of the factor $\frac{1}{2}$, the procedure EDGETRACE accumulates a total equal to twice the enclosed area (AREA2), and divides this by two in the final calculations.

Calculation of first moments of area and centroids

A similar method is used to calculate the first moments of the enclosed area about the axes x = 0 and y = 0. Fig. 11 shows the eight elemental vectors, and the first moments of area of the areas between each vector and the x axis ($\triangle M_x$). The sum of all the $\triangle M_x$ components for a closed outline will yield a negative total which, when divided by the enclosed area, will give a correctly negative value for the y coordinate of the centre of area.

The expression for $\triangle M_x$ relating to odd-numbered vectors contains a constant term, $+\frac{1}{6}$ or $-\frac{1}{6}$. In a closed outline containing a large number of vector elements, each vector may be expected in approximately equal numbers, leading to a cancelling of the constant terms. The constant term will also be small compared with the Y^2 term and the enclosed area, particularly in the case of the close-up view of an object which then largely fills the frame. The calculation of moment of area is therefore simplified by neglecting the constant term. As for the area calculation, the

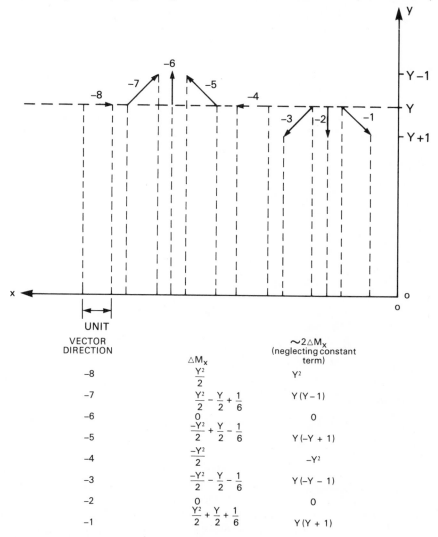

UNIT VECTOR DIRECTION	$\triangle M_x$	$\sim 2\triangle M_x$ (neglecting constant term)
−8	$\dfrac{Y^2}{2}$	Y^2
−7	$\dfrac{Y^2}{2} - \dfrac{Y}{2} + \dfrac{1}{6}$	$Y(Y-1)$
−6	0	0
−5	$\dfrac{-Y^2}{2} + \dfrac{Y}{2} - \dfrac{1}{6}$	$Y(-Y+1)$
−4	$\dfrac{-Y^2}{2}$	$-Y^2$
−3	$\dfrac{-Y^2}{2} - \dfrac{Y}{2} - \dfrac{1}{6}$	$Y(-Y-1)$
−2	0	0
−1	$\dfrac{Y^2}{2} + \dfrac{Y}{2} + \dfrac{1}{6}$	$Y(Y+1)$

Fig. 11. Element contributions for X moment calculation.

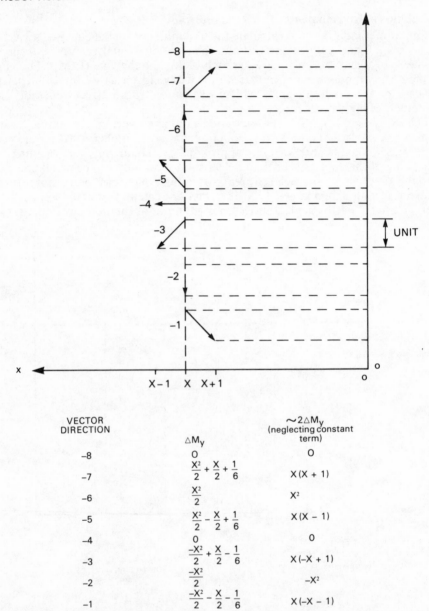

VECTOR DIRECTION	$\triangle M_y$	$\sim 2\triangle M_y$ (neglecting constant term)
−8	0	0
−7	$\dfrac{X^2}{2} + \dfrac{X}{2} + \dfrac{1}{6}$	$X(X+1)$
−6	$\dfrac{X^2}{2}$	X^2
−5	$\dfrac{X^2}{2} - \dfrac{X}{2} + \dfrac{1}{6}$	$X(X-1)$
−4	0	0
−3	$\dfrac{-X^2}{2} + \dfrac{X}{2} - \dfrac{1}{6}$	$X(-X+1)$
−2	$\dfrac{-X^2}{2}$	$-X^2$
−1	$\dfrac{-X^2}{2} - \dfrac{X}{2} - \dfrac{1}{6}$	$X(-X-1)$

Fig. 12. Element contributions for Y moment calculation.

factor $\frac{1}{2}$ is avoided by summing $2\triangle M_x$. The total is accumulated in the double length integer XMOMENT by the procedure SUMXMOMENT.

At the end of the procedure EDGETRACE, the total XMOMENT is divided by AREA2 and the result rounded by giving the centre of area Y coordinate YCENTROID.

Summing moments of area about the y axis gives the value XCENTROID in the same way. The contributions of each vector to this total are shown in Fig. 12.

A single noise point, or a small number of adjacent points representing for example a piece of swarf, will give outlines of a few vectors only. A suitable value for MINCHAIN is found to be −10, that is outlines must have more than ten component vectors to be worthy of further consideration.

The image matrix after EDGETRACE

The edge-following algorithm employed resets each point it visits in the differentiated frame. All the points on an outline are not necessarily visited and consequently reset by the algorithm. Points set to '1' that are then left in the frame will not normally form another complete outline, but they will constitute noise points during subsequent operations and care must be taken to minimise their effect. Figure 13 shows a plot of the image matrix FRAME of Figure 1 and 2 after EDGETRACE has traced both outlines, leaving the residual noise points.

Hole identification – procedure INSIDE

INSIDE determines whether a point located in the image FRAME lies inside or outside an established outline.

It is assumed that the outline generated by EDGETRACE is stored in the array CHAIN with the number of elements in CHAINCOUNT. The starting point of the chain is defined by the integers PARTXSTART, PARTYSTART.

The coordinates of the point to be tested are loaded into STARTX, STARTY and INSIDE is called as a conditional procedure. The 'condition true' return occurs if the test point is inside the stored outline. The stored values are not changed.

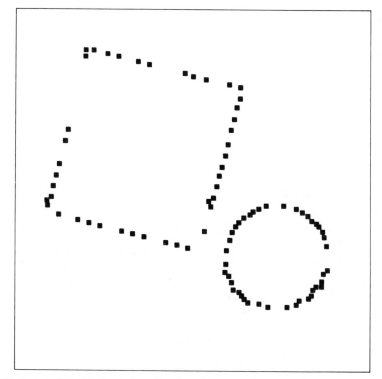

Fig. 13. The frame of Fig. 12 after both outlines have been traced.

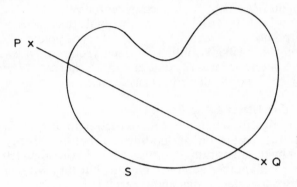

a. Q outside S – PQ intersects S an even number of times.

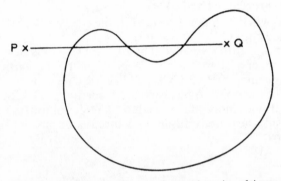

b. Q inside S – PQ intersects S an odd number of times.

Fig. 14. The basis of the INSIDE algorithm.

The algorithm employed is based on the fact that in a plane, if a point P outside a closed curve S is connected by a straight line to another point Q, then Q is also outside S if and only if the line PQ intersects S an even number of times. If PQ intersets S an odd number of times, then Q is inside S (Fig. 14). A proof is given by Courant and Robbins.[14]

In the procedure INSIDE a notional straight line is drawn from the point to be tested to the edge of the FRAME parallel to the x axis, that is the line Y =STARTY, X > STARTX (Fig. 15). A pair of local integers (XCOORD, YCOORD) are initially set equal to the coordinates of the start of the stored outline. The elemental vectors defining the outline are read sequentially from the array CHAIN and used to modify XCOORD, YCOORD (procedure UNCHAIN) so that, XCOORD, YCOORD step through the coordinates of each point traced in the original outline.

At each step original outlines are tested to see if the point lies on the line y = STARTY, x > STARTX. The number of times this line is crossed is counted in the integer CROSSCOUNT. Multiple crossings must not be recorded when the outline lies on the test line for a number of consecutive points (Fig. 16). When the test line is met, a flag word is set, and the vector direction on meeting the line is noted (STRIKEDIRN). For each subsequent outline point lying on the test line, no action is taken. When the test line is left, the current, or leaving direction is compared with

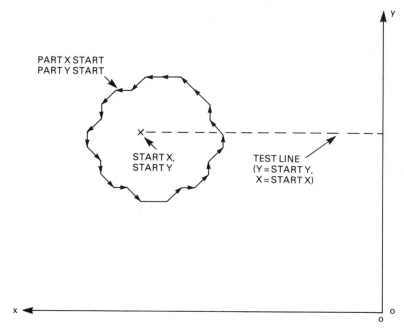

Fig. 15. Illustrating the INSIDE test.

STRIKEDIRN to determine whether the outline has crossed the test line (CROSSCOUNT is incremented) or only run tangentially to it (CROSSCOUNT is not incremented).

When each point represented by the stored outline has been regenerated and tested, CROSSCOUNT is examined and the procedure exists accordingly. The potentially ambiguous case of the tested point lying on the stored outline cannot occur, as the action of tracing the outline resets all the points visited.

It is verified that the object is correct, then its position and orientation must be determined to enable the machine to manipulate it as required by the assembly process.

The problem of orientation

The basic image processing procedures provide an analysis of outlines in the image (including both objects and holes in objects) in terms of area, perimeter, position of centre of area, enclosing rectangle, and chain vector description. These parameters will provide a useful first test, but fuller use of both hole and outline information is required to verify identity and determine orientation.

Fig. 16. The tangent test for procedure INSIDE.

The holes model

The hole pattern in an object can be modified in terms of the distances of the centres of area of the holes from the centre of area of the object, and their relative angular positions. In addition, the areas and perimeters of the holes can readily be measured, yielding the data illustrated in Fig. 17.

This model is independent of the orientation of the parent objects, and can be directly compared with a stored reference model. When the individual holes in the scanned object have been related to the holes in the reference model, the absolute angular positions of the holes in the scanned object can be used to determine the orientation of the object in the input image.

Outline feature extraction

A model of the outline shape is immediately available in the form of the chain vector list. Comparisons of reference models and input data in this form as used for example by Freeman[7] are however complex and time consuming. A major difficulty is that the starting point of the chain on the outline of an object is arbitrary, depending on the orientation of the object and its position in the raster scan. The chains to be compared have therefore to be rotated relative to one

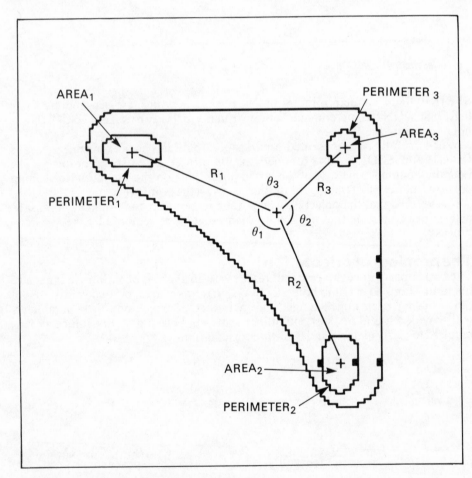

Fig. 17. Illustrating the parameters for the HOLES model.

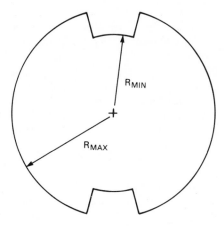

Fig. 18a. Illustrating maximum and minimum radius vectors.

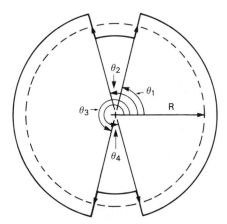

Fig. 18b. Illustrating the parameters for the CIRCLES model.

another whilst searching for a match, and the chain description is not well suited to rotation operations other than in increments of 45°. The correlation is further complicated by the fact that the reference and input are unlikely to have the same number of constituent elements.

A number of techniques have been developed for smoothing chain lists and normalising their lengths to facilitate comparison. Zahn[15] and Montanari[16] described methods of deriving minimal length straight line approximations to reduce the amount of data to be processed. Such techniques are unnecessarily complex in the context of an assembly machine, however. The requirement is to verify the tentative identification of the component, and to determine its orientation rather than to perform 100% inspection by detailed comparison of input and reference outlines.

The need for a model which allows simple, fast matching independent relative orientation suggest that an approach based on feature extraction from the outline may be more appropriate. Many character recognition methods, for example Hosking[17] exploit the fact that the line structure of the characters produces a set of common features such as line starts, joins, reversals and loops. Some corresponding features that could be defined for solid objects are sudden changes in outline direction (corners) and points of maximum and minimum distance of the outline from the centre of area. However, the unconstrained range of shapes that has to be accommodated by the assembly machine limits the usefulness of a technique based on a preprogrammed list of standard features.

A simple, interactive method of feature definition is required, whereby, during the learning phase, points on the outline of an object that are significant either as recognition or orientation criteria, can be specified by a human operator.

The circles model

The angular positions of the maximum and minimum length radius vectors are attractive recognition and orientation criteria, being readily derived from the data generated by EDGETRACE. However, if the length of the radius vector changes only slowly around the maximum and minimum values as the outline is followed, as for example in (Fig. 18a), the positions determined for R_{MAX} and R_{MIN} will be unreliable. This suggests that a more fruitful approach may be to specify a value of radius R, during the learning phase, for which clearly defined vector positions may be determined. In effect a circle of given radius is superimposed on the object, centred on the centre of area, and the intersection of this circle with the outline are defined as feature points. In (Fig. 18b) the feature points are defined by the parameters $R, \theta_1', \theta_2, \theta_3, \theta_4$. The angles may be differenced to produce an orientation invariant set of parameters for identification, and the absolute values used to measure orientation.

The insertion points are determined from the data produced by EDGETRACE by computing the distance of each point on the outline, represented by the chain list, from the centre of area. The points where this distance equals the specified radius are noted, and their angular positions from the centre of area calculated. The intersection points are thus found as an ordered list, the point with which the list

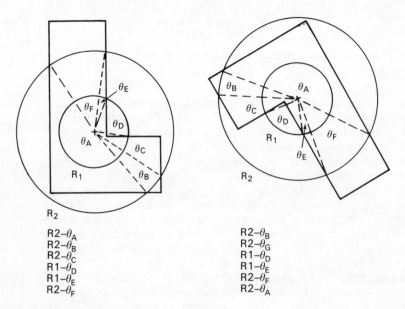

R2–θ_A
R2–θ_B
R2–θ_C
R1–θ_D
R1–θ_E
R2–θ_F

R2–θ_B
R2–θ_G
R1–θ_D
R1–θ_E
R2–θ_F
R2–θ_A

Fig. 19. An example of radius-angle lists for the CIRCLES model.

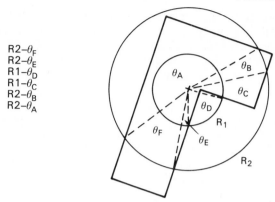

Fig. 20. The component of Fig. 18 shown the wrong way up.

starts being determined by the position on the outline at which EDGETRACE was called. The comparison of the input and reference lists of intersection points may thus require the rotation of one list relative to the other to find a match. Even before the lists are compared, the number of intersections found provides a useful recognition feature.

Suitable criteria for choosing the radius of the circle are:
- the intersection points should uniquely define the orientation of the object,
- the number of intersection points should not be too large say < 8),
- for accuracy in measuring orientation, the radius should be as large as possible,
- the angular positions of the intersection points should not be significantly affected by small changes in radius,
- intersection points should not vanish or new points appear for small changes in radius,

The last two criteria are necessary to minimise the effects of small changes in the input image data.

In order to define uniquely the orientation of a component, or to differentiate it from another similar part, it may be necessary to specify more than one circle. The intersection points found when the outline is traced round will then form an ordered list of radius-angle pairs. In the example shown in Figure 19 two radii are used. The resulting radius-difference angle lists are shown for an object in two different orientations, the list entries being in the order they are found. To compare the two images, the two lists are first checked to ensure that they have the same number of entries. They are then compared for a match in both radius and relative angle entries. If no match is found, one list is rotated and the comparison is repeated. However, since both lists were generated using the same radius data, the radius values may be replaced by integer radius numbers.

The measured angle values need only be compared if an exact match is found between the two lists of radius numbers, thus simplifying the comparison process.

When the correlation between the two lists is established, the difference between the absolute angular positions of each pair of corresponding intersection points can be calculated. The mean of these six values is a reliable measure of the orientation of one object relative to the other. A further property of the circles model is the ability to distinguish the 'wrong way up' situation. Figure 20 shows that with the arrays rotated to give corresponding radius numbers, the angle difference values are in reverse order for the 'wrong way up' component.

The circles model thus provides a powerful method of specifying features for recognition or orientation purposes on the outline of an arbitrarily shaped object.

The amount of storage required for the reference data on one part is only two words per intersection point, and the corresponding data for the scanned part can be derived readily from the results of the EDGETRACE procedure. In conjunction with the holes model, it has therefore been used to provide the recognition and orientation capability of the assembly machine at the lower level.

Concluding remarks

Readers must regard these notes as an introduction only. A great deal has been published on this subject in the last decade – particularly at international conferences. Some of the recent work is included in this collection of papers. Further information on the visually interactive robot SIRCH referred to in this contribution can be recovered from existing published material.[18-22]

References

[1] J. R. Ullman, 'Pattern recognition techniques', Butterworths, London (1972).
[2] A. Rosenfeld, 'Picture processing by computer', Academic Press, New York (1969).
[3] H. G. Barrow and R. J. Popplestone, 'Relational descriptions in picture processing', Machine Intelligence 6, Edinburgh University Press (1971).
[4] H. Freeman, 'On the encoding of arbitrary geometric configurations', *IRE Trans. on Electronic Computers*, June 1961.
[5] H. Freeman, 'Techniques for the digital computer analysis of chain-encoded arbitrary plane curves', *Proc. Nat. Electronics Conf.*, Vol. 17, Chicago (1961).
[6] H. Freeman, 'On the digital computer classification of geometric line patterns', *Proc. Nat. Electronics Conf.*, Vol. 18, Chicago (1962).
[7] H. Freeman, 'A pictorial jigsaw puzzles: the computer solution of a problem in pattern recognition', *IEEE Trans. EC-13*, April 1964.
[8] H. Freeman, 'A review on relevant problems in the processing of line-drawing data', Automatic Interpretation and Classification of Images, Academic Press, New York (1969).
[9] H. Freeman, 'Computer processing of Line-Drawing Images', *Computing Surveys*, Vol. 6, No. 1 (March 1974).
[10] G. P. Dineen, 'Programming pattern recognition', *Proc. Western Joint Computer Conf.*, Los Angeles 1956.
[11] S. H. Unger, 'Pattern detection and recognition', *Proc. IRE*, October 1959, pp. 1737-1752.
[12] A. Rosenfeld, 'Connectivity in digital pictures', *J. Assoc. Computing Machinery*, Vol. 17, No. 1 (January 1970).
[13] P. Saraga and P. R. Wavish, 'Edge coding operators for pattern recognition', *Electronics Letters*, Vol. 7, No. 25 (16 December 1971).
[14] R. Courant and H. Robbins, 'What is mathematics?', Oxford University Press, London (1941) pp. 267-269.
[15] C. T. Zahn, 'A formal description for two-dimensional patterns', *Proc. Joint International Cont. Artificial Intelligence*, Washington 1969.
[16] U. Montanari, 'A note on minimal length polygonal approximation to a digitized contour', *Comm. ACM*, Vol. 13, No. 1 (January 1970).
[17] K. H. Hosking and J. Thompson, 'A feature detection method for optical character recognition', *IEE/NPL Conf. on Pattern Recognition*, London, July 1968.
[18] A. Pugh, W. B. Heginbotham and P. W. Kitchin, 'Visual feedback applied to programmable assembly machines', *Proc. 2nd Int. Symp. on Industrial Robots*, ITTRI, Chicago, May 1972.
[19] W. B. Heginbotham, D. W. Gatehouse, A. Pugh, P. W. Kitchin and C. J. Page, 'The Nottingham "SIRCH" assembly robot', *Proc. 1st Conf. on Industrial Robot Technology*, University of Nottingham, March 1973.
[20] C. J. Page, 'Visual and tactile feedback for the automatic manipulation of engineering parts', Ph.D. Thesis, University of Nottingham, 1974.
[21] C. J. Page and A. Pugh, 'Visually interactive gripping of engineering parts from random orientation', *Digital Systems for Industrial Automation*, Vol. 1, No. 1, pp. 11-44 (1981).
[22] P. W. Kitchin, 'The Application of Visual Feedback to Assembly Machines', Ph.D. Thesis, University of Nottingham, England, October 1977.

RECOGNISING AND LOCATING PARTIALLY VISIBLE OBJECTS: THE LOCAL-FEATURE-FOCUS METHOD

R. C. Bolles and R. A. Cain, SRI International, USA

Reprinted by permission from Robotics Research, Vol. 1 No. 3 –
copyright 1982 The Massachusetts Institute of Technology.

Abstract

A new method of locating partially visible two-dimensional objects is presented. The method is applicable to complex industrial parts that may contain several occurrences of local features, such as holes and corners. The matching process utilises clusters of mutually consistent features to hypothesise objects, also uses templates of the objects to verify these hypotheses. The technique is fast because it concentrates on key features that are automatically selected on the basis of a detailed analysis of CAD-type models of the objects. The automatic analysis applies general-purpose routines for building and analysing representations of clusters of local features that could be used in procedures to select features for other locational strategies. These routines include algorithms to compute the rotational and mirror symmetries of objects in terms of their local features.

Introduction

One of the factors inhibiting the application of industrial automation is the inability to acquire a part from storage and present it to a work station in a known position and orientation. Industrial vision systems that are currently available, like MIC's VS-100 vision system,[1] Bausch and Lomb's OMNICON,[2] and Automatix's vision module[3], can recognise and locate isolated parts against a contrasting background only. These systems recognise binary patterns by measuring global features of regions, such as area, elongation, and perimeter length, and then comparing these values with stored models. Many tasks fit the constraints of these systems quite naturally or can easily be engineered to do so. However, there are also many important tasks in which it is difficult or expensive to arrange for the parts to be isolated and completely visible. In this paper we describe a technique to identify and locate partially visible objects on the basis of two-dimensional models.

The class of tasks that involve the location of partially visible objects ranges from relatively easy jobs, such as locating a single two-dimensional object, to the

The work reported in this paper was supported by the National Science Foundation under Grant No. DAR-8023130.

Fig. 1. Partial view of an aircraft frame member.

extremely difficult task of locating three-dimensional objects jumbled together in a bin. In this paper we concentrate on tasks that are two-dimensional in the sense that the uncertainties in the location of an object are in a plane parallel to the image plane of the camera. This restriction implies a simple one-to-one correspondence between sizes and orientations in the image, on the one hand, and sizes and orientations in the plane of the objects, on the other.

This class of two-dimensional tasks can be partitioned into four sub-classes that are defined in terms of the complexity of the scene:
○ A portion of one object.
○ Two or more objects that may touch one another.
○ Two or more objects that may overlap one another.
○ One or more objects that may be defective.

This list is ordered roughly by the amount of effort required to recognise and locate the objects. Examples of these subclasses are shown in Figs. 1 to 4.

Fig. 1 shows a portion of an aircraft frame member. A typical task might be to locate the pattern of holes for mounting purposes. Since only one frame member is visible at a time, each feature appears at most once, which simplifies feature identification. If several objects can be in view simultaneously and can touch one another (as in Fig. 2), the features may appear several times. Boundary features, such as corners, may not be recognisable, even though they are in the picture, because the objects are in mutual contact. If the objects can lie on top of one another (as in Fig. 3), even some of the internal holes may be unrecognisable because they are partially or completely occluded. And, finally, if the objects are defective (as in Fig. 4), the features are even less predictable and hence harder to find.

Since global features are not computable from a partial view of an object, recognition systems for these more complex tasks are forced to work with either local features, such as small holes and corners, or extended features, such as a large segment of an object's boundary. Both types of features, when found, provide constraints on the positions and orientations of their objects. Extended features are in general computationally more expensive to find, but they provide more information because they tend to be less ambiguous and more precisely located.

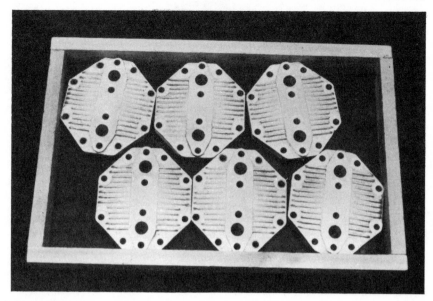

Fig. 2. Mutually touching parts in a tray.

Given a description of an object in terms of its features, the time required to match this description with a set of observed features appears to increase exponentially with the number of features.[4] Therefore, practical applications have been restricted to relatively simple tasks like identifying the holes in Fig. 1. The multiplicity of features in Figs. 2, 3 and 4 precludes the straightforward application of any simple matching technique.

Three approaches have been explored for tasks involving a large number of features. The first is to locate a few extended features instead of many local ones

Fig. 3. Overlapping parts.

Fig. 4. Three parts, one of which is defective.

(e.g., see [5] and [6]). Even though it costs more to locate extended features, the reduction in the combinatorial explosion is often worth it. The second approach is to start by locating just one feature and then use it to restrict the search areas for nearby features (e.g., see [7, 8]). Concentrating on one feature may be risky, but here too, the reduction in the total number of features to be considered is often worth it. The third approach is to sidestep the problem by hypothesising massively parallel computers that can perform matching in linear time. Examples of these approaches include graph-matching,[9, 10] relaxation,[11, 12] and histogram analysis.[6, 13, 14] The advantage of these approaches is that they base their decisions on all the available information.

Although parallel computers surely will be available sometime in the future, in this paper we restrict our attention to methods for sequential computers. The basic principle of the local-feature-focus (LFF) method is to find one feature in an image, referred to as the 'focus feature', and use it to predict a few nearby features to look for. After finding some nearby features, the program uses a graph-matching technique to identify the largest cluster of image features matching a cluster of object features. Since the list of possible object features has been reduced to those near the focus feature, the graph is relatively small and can be analysed efficiently.

The key to the LFF method is an automatic feature-selection procedure that chooses the best focus features and the most useful sets of nearby features. This automatic-programming capability makes possible quick and inexpensive application of the LFF method to new objects. As Fig. 5 shows, the training process, which includes the selection of features, is performed once and the results are used repeatedly.

In this paper we describe the run time system first so as to clarify the requirements of the training system. After describing the training system, we conclude with an evaluation of the LFF method and a discussion of possible extensions.

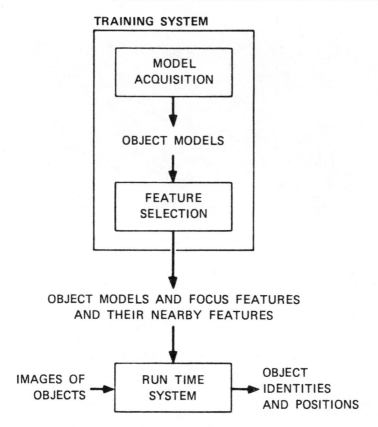

Fig. 5. Top-Level block diagram of the local-feature-focus method.

Run time system

The run time phase of the LFF system acquires images of partially visible objects and determines their identities, positions, and orientations. This processing occurs in four steps:

1. Reading task information.
2. Locating local features.
3. Hypothesising objects.
4. Verifying hypotheses.

The first step, as indicated in Fig. 5, is to input the object models together with the list of focus features and their nearby co-features. Then, for each image the system locates all the potentially useful local features, forms clusters of them to hypothesise object occurrences, and finally performs template matches to verify these hypotheses. In this section we use a simple task to illustrate the basic processing of these steps and then consider a few more difficult tasks that demonstrate additional capabilities of the system. The initial assignment is to locate the object in Fig. 6, which is half of a metal door hinge, in the image in Fig. 7.

The examples in this paper were produced by an implementation of the LFF system in which the run time system is executed on a PDP-11/34 minicomputer and the feature selection runs on a VAX-11/780. The run time software is a modified version of the SRI Vision Module software,[15] which recognises and

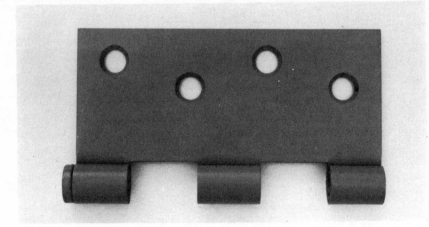

Fig. 6. Half of a metal door hinge.

locates isolated, completely visible objects in binary images. Fig. 8 shows the LFF work station. The camera is a GE TN2500 camera, which provides a 240 × 240 pixel array. It views objects placed directly beneath it on a light table. Grey-scale and binary images are displayed on monitors and the user interacts with the system through a Tektronix display. The fact that the current implementation locates local features in binary images is due to the ready availability of the appropriate software. However, because the techniques for selecting and matching clusters of local features are independent of the manner in which the local features are detected, they can be applied directly to the task of locating objects in grey-scale images or in range data, given local feature detectors for such data.

Fig. 7. Binary image of the hinge part.

Task information

The task information consists of the following:
○ Statistical descriptions of local features.
○ Analytic descriptions of objects.
○ A strategy for locating the objects.

The local features are defined in terms of their appearances in images. The objects and the locational strategy are defined in terms of these local features.

In the current implementation there are two types of features, regions and corners. A region is described in terms of its colour (black or white), area, and axis ratio (the ratio of its minor and major axes). A corner is characterised by the size of its included angle. Each property of a feature is assigned an expected value, and if appropriate, a variance about that value. These variances are important because the system can use them to locate the features efficiently. In particular, the system can use a feature-vector pattern recognition routine, such as the nearest-neighbour algorithm used by the SRI vision module, to identify features.

Although the object descriptions are used mainly at run time to verify hypotheses, it is natural to discuss them before describing the strategies because they contain a list of local features for each object. The features in these lists are referred to as 'object' or 'model' features, as distinguished from 'image' features, i.e., those in images. Each object feature is assigned a unique name and is described in terms of its type and its position and orientation with respect to the object. The positions and orientations have variances associated with them that are designed to model a combination of the imprecision associated with manufacturing the object and the imprecision associated with locating features in an image. The system uses these variances to decide, among other things, whether or not the distance between two image features is sufficiently close to the distance between two corresponding object features to be called a match.

In the current implementation a description of an object also contains a

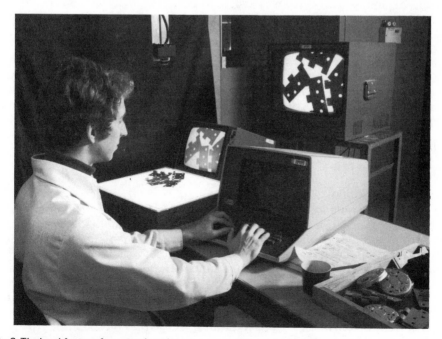

Fig. 8. The local-feature-focus workstation.

description of its boundary. The boundary is represented by a list of points defining a sequence of line segments that approximate the boundary of the object. Once an object has been hypothesised at a particular position and orientation in an image, the boundary is rotated and translated to that location and checked against the image.

The strategy, which is a list of focus features and their nearby co-features, is a reformulation of the object descriptions into a procedure-oriented list of features for finding the objects. For each focus feature there is a list of nearby features, which if found, can be used to identify the focus feature and establish the position and orientation of the object. The description of a nearby feature includes its type, its range of distances from the focus feature, its range of orientations with respect to the focus feature (if it has an inherent orientation), and a list of the object features, anyone of which it might be. Thus, after finding an occurrence of a focus feature, the program simply runs down the appropriate list of nearby features and looks into the image for features that satisfy the specified criteria. As it finds matching features it builds a list of possible object-feature-to-image-feature assignments. This list is transformed into a graph that is analysed by an algorithm for finding the largest completely connected subgraph. The subgraph corresponds to a set of assignments that can be used to hypothesise an object.

Fig. 9 contains the information produced by the feature selection phase for the task of locating the hinge part shown in Fig. 6. Three types of local features are described: holes, A-type corners, and B-type corners. An A-type corner is a concave corner; the hinge occupies threequarters of a small circle centred on the vertex. A B-type corner is a convex corner. The hinge, according to its object description, has four holes, four A-type corners, and eight B-type corners. The object-specific names for the four holes are Hole 1, Hole 2, Hole 3, and Hole 4 (see Fig. 14 for a definition of the object feature names used in Fig. 9).

The best focus feature for locating a hinge, as noted in the strategy, is a hole. The second best focus feature is an A-type corner, the third best a B-type corner. According to the locational strategy, once it has found a hole the system should look for seven nearby features – the first of which is a B-type corner that is between 0.527 and 0.767 inches from the hole and whose orientation with respect to a vector from the hole to the corner is between 155.62 and 180.00 degrees or between −180.00 degrees and −160.03 degrees. If any B-type corners are found that meet these criteria, they are likely to be object feature B6; in this example, by checking all pairs of holes and B-type corners, the training system has narrowed down from eight to one the list of possible object features for this type of corner with respect to a hole. This reduction demonstrates the benefits possible from training-time analysis. In the next few sections we shall use the task information in Fig. 9 to illustrate run time processing.

Local-feature location

The goal of the feature location step, which is the first type of processing applied to a new image, is to find features in the image that match the local-feature descriptions in the task information. The current implementation locates all the features it can and passes a list of them to the hypothesis generation step. The assumption is that in the not-too-distant future there will be special-purpose hardware processors that can locate efficiently all local features in an image. However, if features are relatively expensive to locate, feature detection can be integrated into the hypothesis generation step to minimise processing time.

The current system locates regional features, such as the holes described in Fig. 9, by finding regions in the binary image whose properties are sufficiently close to the

LOCAL-FEATURE DESCRIPTIONS

HOLE	(type:	REGION)	
	(colour	WHITE)	
	(area:	0.071	< variance .001 >)
	(axisRatio:	1.000	< variance .082 >)

A	(type:	CORNER)	
	(chordLength:	12)	
	(includeAngle:	270.0	< variance 40.703 >)
B	(type:	CORNER)	
	(chordLength:	12)	
	(includeAngle:	90.0	< variance 40.703 >)

OBJECT DESCRIPTION

HINGE	(boundrayList:	(0.000,0.000,	0.000,4.000,	2.230,4.000,
		2.230,3.310,	1.700,3.310,	1.700,2.430,
		2.230,2.430,	2.230,1.625,	1.700,1.625,
		1.700,0.745,	2.230,0.745,	2.230,–0.13,
		1.700,–0.13,	1.700,0.000,	0.000,0.000))

(objectFeature	type	(x,y)	Orient	xyVar	oriVar:
(Hole1	HOLE	0.440,0.475	—	0.004	0)
(Hole2	HOLE	0.790,1.475	—	0.004	0)
(Hole3	HOLE	0.440,2.475	—	0.004	0)
(Hole4	HOLE	0.790,3.475	—	0.004	0)
(a1	A	1.700,1.625	135.0	0.010	21.447)
(a2	A	1.700,0.745	–135.0	0.010	21.447)
(a3	A	1.700,2.430	–135.0	0.010	21.447)
(a4	A	1.700,3.310	135.0	0.010	21.447)
(b1	B	0.000,4.000	–45.0	0.010	21.447)
(b2	B	2.230,4.000	–135.0	0.010	21.447)
(b3	B	2.230,3.310	135.0	0.010	21.447)
(b4	B	2.230,2.430	–135.0	0.010	21.447)
(b5	B	2.230,–0.13	135.0	0.010	21.447)
(b6	B	0.000,0.000	45.0	0.010	21.447)
(b7	B	2.230,0.745	–135.0	0.010	21.447)
(b8	B	2.230,1.625	135.0	0.010	21.447)

FOCUS FEATURES AND THEIR NEARBY COFEATURES

HOLE
(possibleObjectFeatures: (Hole1 Hole2 Hole3 Hole4))

(nearbyFeat	distanceRange	orientationRange	possibleObjectFeatures:
(B	0.527,0.767	155.62,–160.03	b6)
(A	0.801,1.041	107.32, 145.25	a1)
(B	0.828,1.068	149.83,–172.63	b1)
(HOLE	1.019,1.099	— , —	Hole1, Hole2, Hole3, Hole4)
(A	1.140,1.380	–149.83,–116.02	a2 a3)
(B	1.467,1.707	–166.85,–135.28	b1 b2)
(A	1.050,1.290	–113.39, –78.67	a2))

A
```
(possibleObjectFeatures:  (a1 a2 a3 a4))
(nearbyFeat          distanceRange        orientationRange  possibleObjectFeatures:
    (B               0.325,0.725          100.21, 167.93           b3 b8         ))
    (B               0.325,0.725         −168.62,−100.96           b4 b7          )
    (B               0.820,1.220          171.27,−143.20           b2 b5 b8       )
    (A               0.610,1.010          109.21, 160.72           a3             )
    (HOLE            1.391,1.631             —   ,   —             Hole3          )
    (A               0.610,1.010         −160.37,−108.86           a1             )
    (HOLE            1.050,1.290             —   ,   —             Hole2 Hole3    )
    (HOLE            1.173,1.413             —   ,   —  Hole1 Hole2 Hole3 Hole4   )
```

B
```
(possibleObjectFeatures:  (b1 b2 b3 b4 b5 b6 b7 b8))
(nearbyFeat          distanceRange        orientationRange  possibleObjectFeatures:
    (B               0.490,0.890         −163.29,−106.69           b3 b5 b8      ))
    (A               0.670,1.070         −122.21, −72.71           a1 a4          )
    (HOLE            0.828,1.068             —   ,   —             Hole4          )
    (B               0.680,1.080          −69.56, −20.40           b4 b7          )
    (B               0.490,0.890          109.84, 159.34           b2 b4 b7       )
    (A               0.820,1.220           81.59, 127.14           a2 a3          )
    (HOLE            0.527,0.767             —   ,   —             Hole1          )
    (HOLE            1.467,1.707             —   ,   —             Hole2 Hole3
                                                                     Hole4)

    (A               0.325,0.725           12.03,  79.76           a2 a3          )
    (B               1.490,1.890          −63.60, −27.10           b4 b7          )
```

Fig. 9. Task information for the hinge part.

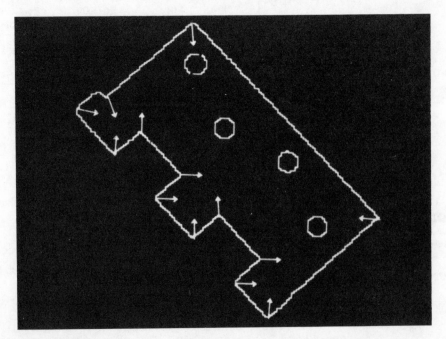

Fig. 10. Corners detected in Figure 7.

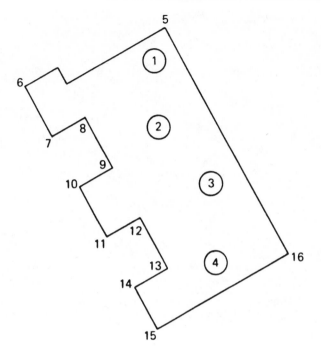

Fig. 11. Image feature numbers.

nominal values. The system locates corners by moving a jointed pair of chords around the boundaries and comparing the angle between the chords to the angles defining the different types of corners. Fig. 10 shows the corners located in the image in Fig. 7. This method of finding corners is only one of many possible methods. It was chosen for its simplicity and speed. It encounters difficulties with rounded corners and its precision is influenced by image quantisation, but we have found it to be an effective way of finding corners.

The product of the feature location step is a list of local features found in the image. Given the image feature numeration in Fig. 11, the list for Fig. 10 is shown in Fig. 12. The program at this stage in the processing has not yet determined the object feature names – it has just determined their types (i.e., hole, A-type corner, or B-type corner). To assign object feature names the program has to analyse the relative positions and orientations of the features. This processing is described in the next subsection.

1. hole	2. hole	3. hole	4. hole
5. B	6. B	7. B	8. A
9. A	10. B	11. B	12. A
13. A	14. B	15. B	16. B

Fig. 12. List of local features found in Figure 7 and their types.

Hypothesis Generation

The goal of the hypothesis generation step is to generate good hypotheses as fast as possible. As usual there is a trade-off between good and fast. The optimum procedure is a function of the cost of verifying a hypothesis relative to the cost of

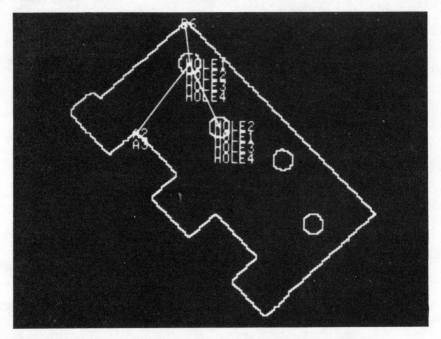

Fig. 13. Nearby features found around a hole and their lists of the possible object features.

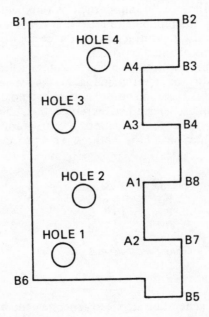

Fig. 14. Local features of a hinge part.

Hole1-to-image-feature-1   Hole2-to-image-feature-1
Hole3-to-image-feature-1   Hole4-to-image-feature-1
Hole1-to-image-feature-2   Hole2-to-image-feature-2
Hole3-to-image-feature-2   Hole4-to-image-feature-2
B6-to-image-feature-5   A2-to-image-feature-8
A3-to-image-feature-8

Fig. 15. List of object-feature-to-image-feature assignments.

generating one. The LFF system has been developed under the premise that the cost of adequate verification is too large to be ignored. Therefore, it is important to generate the best hypotheses possible.

The run time system hypothesises objects by recognising clusters of image features that match clusters of object features. To find these clusters and avoid as much of the combinatorial explosion as possible the system locates one feature around which it tries to 'grow' a cluster. If this does not lead to a hypothesis, the system seeks another focus feature for a renewed attempt.

In our example of the hinge, the feature selection step rated holes as the best focus features. Consequently, the system looks first for holes. For the image in Fig. 7 the system starts with the uppermost hole. Having selected one, the system then searches the list of local features for those that fit the specifications for features near holes. Fig. 13 shows the nearby features found around the hole. Beside each feature is a list of the object features it could be (see Fig. 14 for definitions of the object features). Fig. 15 lists this information in terms of possible object-feature-to-image-feature assignments.

Given the list of possible assignments in Fig. 15, the run time system uses a graph-matching technique to locate the largest clusters of mutually consistent assignments. In the current implementation, the technique being used is a maximal-clique algorithm (see appendix). To apply this technique, the system transforms the list in Fig. 15 into the graph structure in Fig. 16. Each node in the graph represents a possible assignment of an object feature to an image feature. Two nodes in the graph are connected by an arc if the two assignments they represent are mutually consistent. To be mutually consistent, a pair of assignments must meet the following criteria:

○ The two object features must not refer to the same image feature.
○ The two image features must not refer to the same object feature.
○ The two image features must refer to object features that are part of the same object.
○ The distance between the two image features must be approximately the same as the distance between the two object features.
○ The relative orientations of the two image features with respect to the line joining their centres must be approximately the same as the relative orientations of the two object features.

The assignment Hole 1-to-image-feature-1 is mutually compatible with Hole 2-to-image-feature-2, because image feature 1 is approximately the same distance from image feature 2 as object feature Hole 1 is from object feature Hole 2 in the analytic model. Hole-1-to-image-feature-1 is not mutually compatible with Hole 2-to-image-feature-1, because both object features are assigned to the same image feature. Similarly, Hole 1-to-image-feature-1 is not mutually compatible with Hole 1-to-image-feature-2, because one object feature cannot be assigned to two different image features.

Once the graph has been constructed, the graph-matching technique is used to

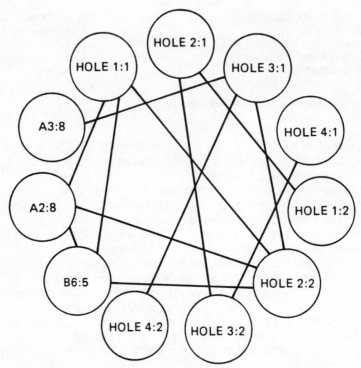

Fig. 16. Graph of pairwise consistent assignments.

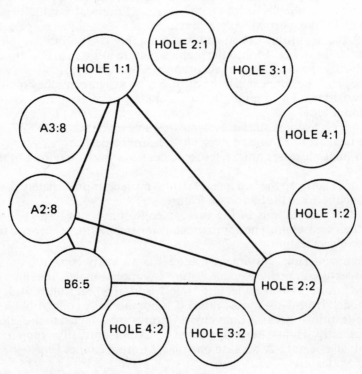

Fig. 17. Largest maximal clique for the graph in Figure 16.

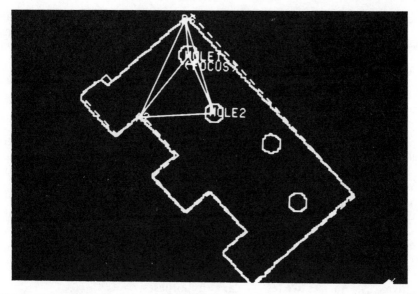

Fig. 18. Local feature identities and a hypothesised hinge location.

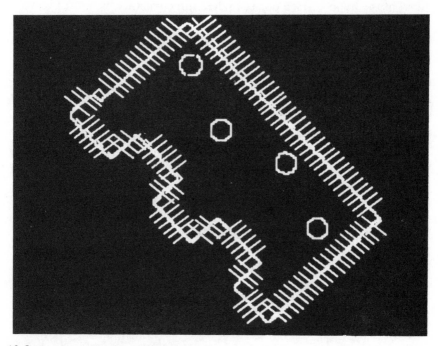

Fig. 19. Segments examined to verify hinge boundary.

extract the largest sets of mutually consistent clusters of nodes. The largest completely connected subgraph (i.e., the largest maximal clique) for the hinge is shown in Fig. 17. It contains four nodes representing the following assignments:

> Hole 1 to image feature 1
> Hole 2 to image feature 2
> A2 to image feature 8
> B6 to image feature 5.

From these assignments the system can compute the translational and rotational offsets required to align the analytic object with this cluster of image features. Fig. 18 shows the assignments of object features to image features and a dashed line indicating the location of the hinge implied by these assignments.

Occasionally there is no unique 'largest' completely connected subgraph. In that case each subgraph is used to form an hypothesis. Thus, the analysis of a graph can produce more than one hypothesis. It is the responsibility of the next phase of the process to determine which, if any, of these hypotheses are valid.

Hypothesis Verification

The current system uses two tests to verify hypotheses:
- It looks at the image for other object features that are consistent with the hypothesis.
- It checks the boundary of the hypothesised object.

As the program finds image features that match predicted object features, it adds them to a list of verified features (which adds strength to the initial hypothesis). It also uses the verified features to improve its estimate of the position and orientation of the object.

Given the refined estimate of an object's location, the program checks the boundary of the hypothesised object to determine whether or not it is consistent

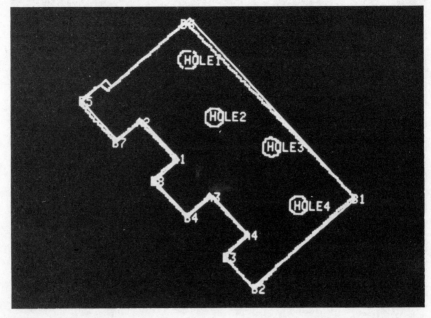

Fig. 20. Verified hinge and its feature labels.

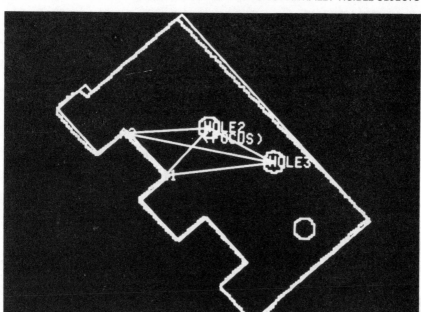

Fig. 21. Cluster of features used to locate the hinge when one hole is covered up.

with the image. To do this, the system rotates and translates the analytic boundary and analyses the contents of the image along the boundary. It takes samples perpendicularly to the boundary and checks for dark-to-light transitions (see Fig. 19). Light-to-dark transitions and all-light samples are negative evidence. All-dark samples are neutral.

If a sufficient number of object features is located and the boundary verified, the system reports the identity of the object, its location, and orientation. As information to be utilised in higher-level processing, it also reports how many features it was able to locate, the number it expected to see in the field of view, and the total number of features expected for the object. Finally, it marks the features as 'explained' in the list of image features, so that additional processing does not attempt to re-explain them. Fig. 20 shows the hinge with all its features labelled.

When a cluster of features leads to two or more hypotheses, typically only one of them passes the verification tests. However, if two or more competing hypotheses pass the tests, the image is ambiguous and the system signals the problem by declaring each hypothesis as a 'possible' match. Additional data are required to disambiguate such cases.

This completes the discussion of the three basic steps in the run time system: local-feature location, hypothesis formation, and hypothesis verification. In the next subsection we consider the application of these steps to more complex tasks.

Examples and Extensions

To demonstrate the full capability of the run time system, we consider a sequence of increasingly difficult tasks. For example, if the hole that was used as the focus feature in the previous task is covered up, the system is forced to select a different focus feature – in this instance another hole. A completely different group of nearby features is found (see Fig. 21) that matches a different cluster of object features. However, the implied hypothesis is the same.

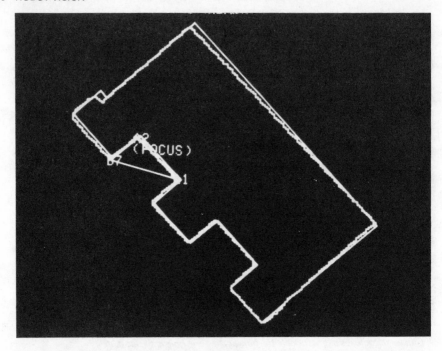

Fig. 22. Cluster of features used to locate the hinge when all the holes are covered up.

Fig. 23. Image of four overlapping hinge parts.

If all the holes have been covered, the system is forced to use the next type of focus feature – in this case, an A-type corner. Once again, a completely different cluster of features is found (see Fig. 22). This cluster leads to two hypotheses, one for each of the rectangular intrusions into the part. The verification step selected the correct match.

In a more complex scene, such as Fig. 23, the system uses many different clusters of local features to locate the objects. This figure shows partially obscured hinges that are touching and overlapping. Using the techniques described above, the system is able to identify and locate all four hinges (see Fig. 24). With the current implementation this processing takes approximately eight seconds.

The run time system also has the ability to recognise and locate objects that are upside down in the image. There is an operator-selectable option that invokes special decision-making procedures to cope with that possibility. If this option is selected, the run time system does the following:

☐ When nearby features are found, they are allowed two orientations; the one specified in the object file and its mirror image.

☐ When the 'mutually compatible' check is made, mirror-image orientations are allowed.

☐ Once a hypothesis has been made, the system determines whether or not the object is upside down by analysing three image features and their corresponding object features. Using any one of the three as a vertex, an included angle is computed for the image features and another one for the object features. If these included angles differ in sign, the object is upside down.

☐ If the system determines that an object is upside down, it performs all subsequent processing with a mirror image of the analytic object (i.e., feature verification and boundary checking).

Figure 25 contains five hinges. The run time system is able to identify and locate all five of them (see Fig. 26), even though three are upside down. This processing

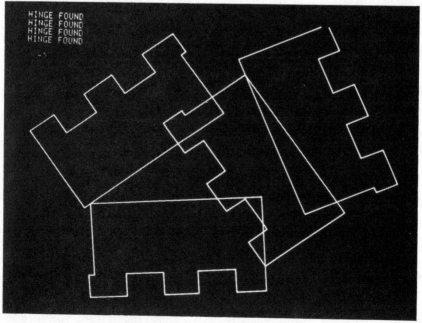

Fig. 24. Hinge parts located in Figure 23.

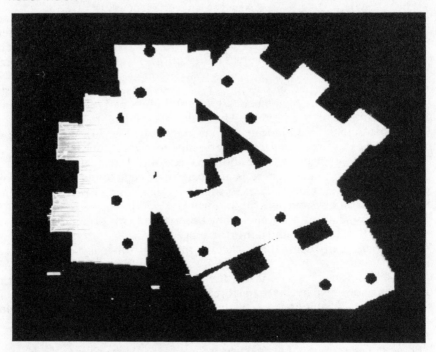

Fig. 25. Image of five hinges, some of which are upside down

takes about 25 seconds in the current implementation. The main reason for the increase in processing time for this example over the previous example (shown in Figs. 23 and 24) is that the system has to distinguish between four similar objects: the two-pronged hinge part, the three-pronged hinge part, and the mirror images of these parts. The additional possibilities increase the sizes of the graphs to be analysed, which increase the processing times.

Training system

The primary design goal for the training system is to make it as easy as possible to use. There are basically two ways to do this. One is to develop automatic techniques for producing as much of the information as possible and the other is to human-engineer the system as well as possible. We have concentrated on the first approach because we are interested in developing ways to capitalise on the information that is commonly available for industrial tasks, such as descriptions of objects and constraints on viewing conditions. In this section we describe techniques for analysing clusters of local features and show how they can be used to produce the information required by the LFF run time system. Since these techniques include algorithms for building and analysing representations of the geometry of local features, they could also be used to implement automatic training systems for other locational strategies based on local features, such as histogramming the orientations suggested by pairs of local features.

The LFF training system is divided into two major parts: model acquisition and feature selection. This model acquisition step produces models of objects to be recognised, while the feature selection step chooses the best features for the LFF recognition procedure.

Model Acquisition

The purpose of the model acquisition step is to construct models of the objects to be recognised. As presented earlier, these models include descriptions of the local-feature types and a list of local features associated with each object. There are several ways to construct these models. There has been a tradition of 'teaching by showing', in which a user shows the system several examples of an object and the system gathers statistics that are used to estimate the expected values of the object's global features and their variances. This approach is straightforward. It can be tedious, however, because a large number of examples is required to produce valid statistical models. Therefore, we are exploring alternative teaching methods that avoid the need for multiple examples.

There are two steps in our approach to defining an object model. The first is specification of the nominal positions and appearances of the local features. The second is estimation of the variances associated with these positions and properties. In the current system, the user can either use a computer-aided-design (CAD) model of an object to specify the nominal values or interactively point out features in an image of the object. The system then uses analytic and probabilistic models of the quantisation errors to predict the variances about these values. The variances for a region's position and area are computed by means of formulas developed by John Hill.[16] Variances for the other properties, such as a region's orientation or a corner's position, are estimated by a Monte Carlo technique that perturbs them in accordance with the predicted size of an individual pixel.

We have tuned the current system so that it is conservative in its estimates of variances. That is, any errors in its estimates are overestimates. Slightly inflated variances lead to somewhat longer recognition times because more feature assignments are consequently made and analysed. On the other hand, underestimates

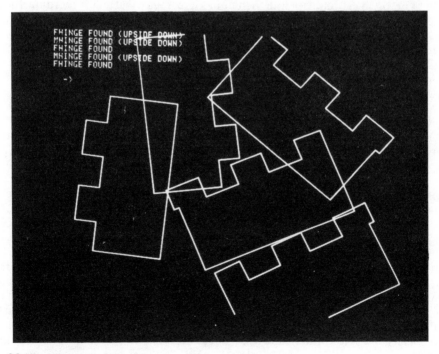

Fig. 26. Hinges located in Figure 25.

Fig. 27. A round electric box cover and a sheet metal part.

could cause the system to miss marginal objects. In the event, the user can adapt the system to his specific needs.

Feature Selection

Given a set of object models, the feature selection process chooses key features and key clusters of features for the LFF method. It selects a set of nearby features for each type of focus feature and then ranks the focus features according to the predicted costs of using them to recognise and locate objects. This selection and ranking are performed as follows:

○ Identify similar local features in different objects.
○ Compute symmetries of the objects.
○ Mark structurally equivalent features.
○ Build feature-centred descriptions of the objects.
○ Select nearby features.
○ Rank the focus features.

Only the last two steps are specifically tailored to the LFF method. The other steps could serve in the preliminary analysis for other locational strategies. Each step is described in a separate subsection.

Similar local-feature types. Each object model contains descriptions of the types of local features associated with that object. The purpose of the first step in the feature selection process is to determine the similarity of these features and produce a single list of local-feature types for the set of objects being considered. Identifying similar features is important, because the system needs to determine which features can be reliably distinguished on the basis of local information and which cannot. Or, put another way, the system needs to identify those features that can be distinguished only by the clusters of features near them. Consider the two parts in Fig. 27. Each of them contains three types of holes. Both parts contain eighth-inch and quarter-inch round holes. Therefore, if the system is asked to recognise these two objects, it would combine the two lists of hole types into one list containing four types.

The identification of similar features is straightforward when the ranges of the property values describing the features are either disjoint or almost identical. The difficulty arises when several feature descriptions overlap in complex ways. The current system groups features together if the ranges of their property values (defined by the expected values and associated variances) overlap significantly. The user can specify the amount of overlap to be considered significant and can modify the groups suggested by the system. Reliance on the user for the final decision is a basis principle in our design of a semi-automatic programming system. The system automates as much of the process as possible and displays the results for the user's approval.

The incorrect grouping of features can lead to inefficient locational strategies. To group features that are dissimilar forces the run time system to use more nearby features than necessary. The additional nearby features increase the sizes of the graphs to be analysed, which in turn increase processing times. The source of the inefficiency is the use of nearby features to make distinctions that could be made reliably on the basis of the features' local appearances.

Not grouping features that are locally ambiguous can also lead to inefficient strategies. In this case the feature selection system is unaware of ambiguities to be resolved – and hence selects sets of nearby features that are not designed to make these distinctions. As a result, the run time system is forced to proceed sequentially through possible interpretations of a focus feature until a verifiable hypothesis has been achieved. Since progressing through interpretations is time-consuming, it should be minimised. Thus, it is inefficient to form groups that are either too small or too large. Determining the best groups is difficult because it is a complex function of the cost of generating and verifying hypotheses. The current LFF training system forms reasonable groups for moderately difficult tasks and provides a convenient way for the user to experiment with variations of its suggestions for more difficult tasks.

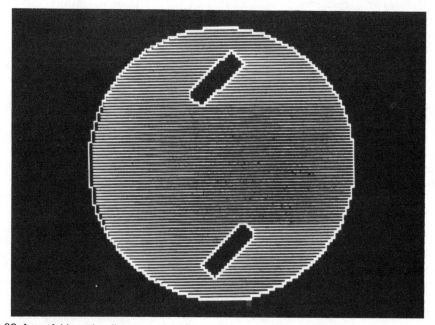

Fig. 28. A two-fold rotationally symmetric object.

Symmetries of objects. Two fundamental properties of two-dimensional objects are their rotational and mirror symmetries. These properties are important in the recognition and location of objects because they can be used to identify key features that determine the orientations of the objects and differentiate them from those that are similar. These properties are especially important in performing industrial vision tasks, in which symmetrical, or almost symmetrical, objects are common. For example, the round part in Fig. 27 is two-fold rotationally symmetric (i.e., its appearance is unchanged by a rotation of 180 degrees). If it occasionally occurs upside down in the scene, it would be important to know that it is not mirror-symmetric. Since it is not mirror-symmetric, a vision system could detect upside-down parts.

A few artificial intelligence programs have performed symmetry analysis. Evans' program, which worked geometric-analogy problems, tested the primitive figures to see whether they were mirror-symmetric about a horizontal or vertical axis.[17] Gips' program analysed two three-dimensional strings of cubes to determine whether they were rotational or mirror transformations of each other.[18] Perkins' program[5] used a form of correlation to determine whether a pattern was rotationally symmetric. The latter program, like its counterparts, was not intended to perform a general-purpose symmetry analysis aimed at understanding local features, clusters of local features, and their properties. It is precisely this kind of comprehensive understanding of objects that the algorithms in this section are being developed for.

Kanade[19] has investigated skewed symmetry in images and its relationship to the gradient space of surface orientations. He uses skewed symmetry to hypothesise real symmetry in a three-dimensional scene and to constrain the associated surfaces. Wechsler[20] describes an algorithm that decomposes two-dimensional regions into mirror-symmetric components. His approach employs mathematical descriptions and tests for symmetry that are similar to the ones described in this section, but his goal is to describe regions rather than characterise clusters of local features. Silva[21] describes a program that derives rotational symmetry axes for three-dimensional objects from multiple views of the objects. His task is more difficult than the problem discussed in this section because he is only given a set of images of the object, not an analytic model of it.

In the LFF system local features are characterised by the following attributes:

O Type (e.g., 90-degree corner or quarter-inch hole).
O x-y position in the object's coordinate system.
O Orientation, if any, with respect to the x axis of the object's coordinate system.
O Rotational symmetry.

Round holes do not have an inherent orientation. By convention, the orientation of a corner is the orientation of the bisector of the angle pointing into the object's interior. Rectangular holes, such as the ones in Fig. 28, are two-fold rotationally symmetric and are assigned a nominal orientation along the major axis of the rectangle.

The position and orientation of a feature with respect to the centroid of the object are important quantities for symmetry analysis. They are defined in terms of the vector from the centroid of the object to the centre of the feature. We refer to this vector as the 'central vector' for the feature. The angular position of the feature is defined to be the counter-clockwise angle from the x-axis of the object to its central vector. The distance of the feature from the centroid is the magnitude of the central vector. The relative orientation of the feature is the smallest counter-clockwise angle from the central vector to one of the feature's axes of symmetry (see Fig. 29). The relative position and orientation of one feature with respect to another is

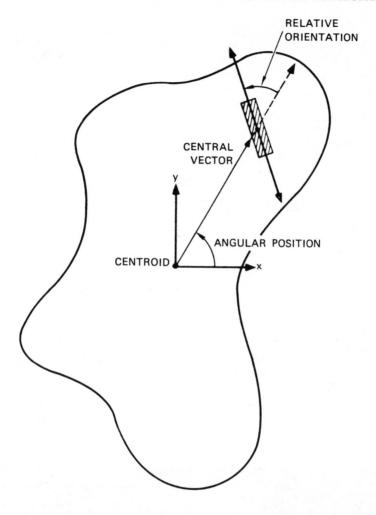

Fig. 29. Diagram of the relative position and orientation of a feature with respect to an object.

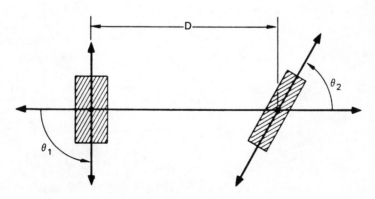

Fig. 30. Diagram of the relative position and orientations of one feature with respect to another.

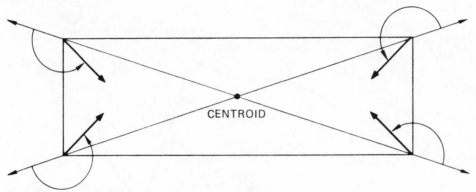

Fig. 31. Relative orientations of the corners of a rectangle.

defined in terms of the three factors, D, θ_1, and θ_2, illustrated in Fig. 30.

The LFF system determines the symmetries of a two-dimensional object in three steps: (1) formation of groups of similar features that are equidistant from the centroid of gravity of the object; (2) computation of the symmetries of the individual groups; (3) computation of the object's symmetries in terms of those of the groups.

Features are defined to be similar for the purposes of group formation if they are the same type, are equidistant from the object's centroid, and share the same orientation relative to their central vectors. Therefore, if two features are similar, the object can be rotated about its centroid so that one feature will be repositioned at the other feature's original position and orientation. For example, the four corners of a square are similar. However, only the diagonally opposite corners of a rectangle are similar, because the relative orientations of two adjacent corners are different (see Fig. 31). Therefore, in forming groups for symmetry analysis, the system divides the four corners of a rectangle into two groups.

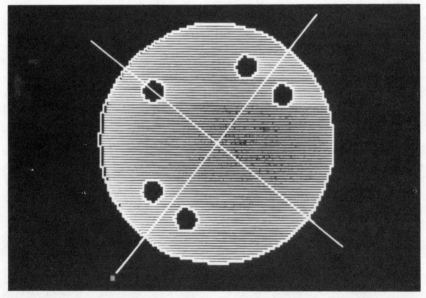

Fig. 32. An object, its axis of least second moment, and the axis perpendicular to that axis.

The second step in the symmetry analysis computes the rotational and mirror symmetries of each group. If a group of N features is D-fold rotationally symmetric, then D must be a divisor of N. Therefore, to determine the rotational symmetry of a group, it suffices to produce the list of factors of N and test the group for D-fold symmetry for each factor, the largest first. To test for D-fold symmetry, the system adds 360/D degrees to each of the angular positions of the features and compares the new positions with the original. If all the new positions are within a specified tolerance of the original ones, the group is D-fold rotationally symmetric.

If a set of two-dimensional points is mirror-symmetric, there is an interesting relationship between the axis of symmetry and the axis of least second moment: they are either identical or perpendicular to each other.[22] Therefore, to test a set of points for mirror symmetry, it is not necessary to do so in all possible orientations; it is necessary only to compute the axis of minimum second moment and to test the set for mirror symmetry about that axis and the one perpendicular to it. Consider the five holes in Fig. 32, which are mirror-symmetric. The axis of second moment is the axis that points to the upper right corner. The pattern is not mirror-symmetric about the latter axis, but it is mirror-symmetric about the axis perpendicular to it.

For a group of features produced by the first step in the symmetry analysis to be mirror-symmetric, their centre must form a mirror-symmetric pattern and each separate feature must be mirror-symmetric about its central vector. The pair of slots in Fig. 28 is not mirror-symmetric, because the individual features are not symmetric about their central vectors. To test a group for mirror symmetry, the system first checks one feature for mirror symmetry about its central vector and then checks the pattern of the centres for mirror symmetry.

It is interesting to note that a group of features, such as the group of five holes in Fig. 32, can be mirror-symmetric and not rotationally symmetric. It is also possible for a group of features to be rotationally symmetric, but not mirror-symmetric (e.g., consider the pair of slots in Fig. 28).

The third step of the symmetry analysis computes an object's symmetries in terms of those of the groups. The rotational symmetry of the object is easy to compute. It is the greatest common divisor of the symmetries of the groups.[22] The mirror symmetry and associated axes of mirror symmetry are more complicated to determine. Once the rotational and mirror symmetries of the groups are known, it is possible to compute a list of mirror symmetry axes for each group. The basic idea is to construct these lists and intersect them to produce the mirror symmetry axes of the object. However, there is one special case to be considered. This case is illustrated by the four corners of a rectangle. As was indicated above, they are divided into two groups according to their orientations relative to their central vectors. Neither group is mirror-symmetric, because the features are not mirror-symmetric about their central vectors. But the set of four corners, taken as a single group, is mirror-symmetric. The problem is that groups can occur in conjugate pairs. Therefore, to test an object for mirror symmetry, the system locates mirror-symmetric groups and pairs of groups that form mirror-symmetric patterns conjointly. If all the features of an object occur in mirror-symmetric patterns and there exists at least one common mirror-symmetry axis, the object is mirror symmetric.

We have implemented these tests and confirmed that they work well within the LFF system. We also tried to incorporate them in a system to build models automatically, for a large class of two-dimensional objects, from images of the objects. We found this goal hard to achieve, as it was difficult to formulate a general definition of 'significant feature'.

Structurally equivalent features. If an object is rotationally symmetric, its features occur in groups whose members are structurally equivalent. That is, they cannot be distinguished on the basis of their local appearance or on the basis of the relative positions and orientations of other features in the object. For example, since the round object in Fig. 27 is twofold rotationally symmetric, its features occur in pairs whose members are structurally equivalent. The implication of structural equivalance for the LFF method is that only one member of a group of structurally equivalent features has to be analysed. Therefore, if an object is rotationally symmetric, the number of features to be considered can be reduced by a factor equal to its rotational symmetry. Any reduction in the number of features is important because of the combinatorial aspects of the matching problem. A reduction in the number of features by a factor of two or more is quite significant.

Having determined the symmetry of an object, the system marks as duplicates those features that are not to be regarded as focus features. Our initial marking procedure arbitrarily divided the features into rotational units, according to their angular positions with respect to the object's x-axis, and marked all features except those in the first unit. For example, for an object that was twofold rotationally symmetric, all features whose angular positions were between 180 degrees and 360 degrees were marked as duplicates. However, since the LFF system concentrates on local clusters of features, an improved marking procedure was developed that locates the most compact cluster of features constituting a complete rotational unit. For the round object in Fig. 27, this procedure divided the features into top and bottom halves, because the features are more closely grouped that way than in another kind of partition. This new strategy maximises the intersection of the sets of nearby features (relative to a focus feature), which in turn minimises the number of possible interpretations of the nearby features.

Feature-centred descriptions. The next step after marking the duplicate features is to build feature-centred descriptions of the objects for each unique feature. These descriptions are designed to encode the information necessary for selecting nearby features. They are essentially the same as the descriptions generated by rotational symmetry analysis, except that they are centred on a local feature instead of the centroid of an object; in addition, the orientation of that feature, if any, is used in the formation of the groups.

The program builds a description for each non-duplicate feature. It does this by partitioning all features of the object, whether they have been marked as duplicates or not, into groups – and then computing the rotational symmetries of the groups. Features are grouped together if they are equidistant from the focus feature, are at the same orientation with respect to vectors from the focus feature, and if the focus feature is at the same orientation with respect to vectors from the features to be grouped (see Fig. 30 for these relative angles). If two features are grouped together, they cannot be distinguished by their relative positions or orientations with respect to the focus feature. Looked at another way, if two features are in the same group, the structural unit composed of the focus feature and one of those features is identical to the structure formed by the focus feature and the other non-focus feature. Fig. 33 illustrates the grouping of features around a hole in the hinge part.

Nearby feature selection. Given feature-centred descriptions of the objects to be recognised, the nearby-feature selection step chooses a set of nearby features for each type of focus feature. It uses the feature-centred descriptions to suggest nearby features and evaluate them by locating matching features near other occurrences of the focus feature. As described in the discussion of the LFF task information a nearby feature is not really a feature; it is rather a set of criteria describing a class of

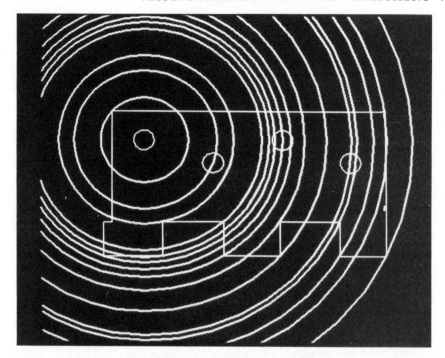

Fig. 33. Feature-centred description of the hinge with respect to one of its holes.

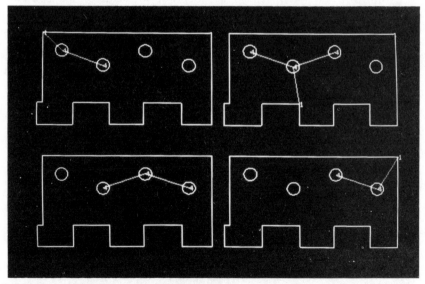

Fig. 34. A unique cluster of nearby features for each hole.

features with respect to another feature. Therefore, in an image or in an object model there may be zero or more features that fit these criteria in the vicinity of a particular focus feature. The selection procedure creates descriptions of nearby features from examples of features in the models.

The basic requirement for a set of nearby features with respect to a focus feature is that the set contain enough features to identify the structurally different occurrences of the focus feature and establish the position and orientation of the object with respect to that feature, In other words, having found the features that match the nearby-feature descriptions, the system should be able to identify the focus feature and all nearby features, then compute the position and orientation of the object. For example, Fig. 34 shows four clusters of features that are sufficiently different to enable the hole they are centred upon to be identified and the object's position and orientation computed.

The current strategy for selecting nearby features is a two-step process:
1. Select nearby features to identify the focus feature.
2. Select additional features, if necessary, to determine the position and orientation of the object.

Features used to identify the focus feature are chosen first because the selected features amost always determine the position and orientation of the object as well. Features selected to determine the position and orientation often do not identify the focus feature.

When an additional feature is needed, those closest to the focus feature are considered first. The assumption is that features close to an observed feature are the most likely to be visible, both because they will probably be in the image and because they are less likely to be occluded.

To illustrate the selection process, let us consider the task of choosing nearby features for the hinge part shown in Fig. 14. There are three types of local features and hence three possible focus features – holes, A-type corners, and B-type corners. Let us start by selecting features for a hole-type focus feature. The first goal is to select nearby features that will allow us to identify which of the four holes has been found. To do this, the system selects features that differentiate between pairs of possible interpretations. That is, it chooses features to distinguish Hole 1 from Hole 2, Hole 1 from Hole 3, and so on. To distinguish Hole 1 from Hole 2, it selects corner B6 with respect to Hole 1 (see Figs. 14 and 35). Corner B6 is the closest feature to either Hole 1 or Hole 2 and can be used to distinguish Hole 1 from Hole 2 because the latter doesn't have a B-type corner at that distance. In fact, since Hole 1

Fig. 35. Distance ranges for B-type corners with respect to a hole.

is the only hole to have a B-type corner near it like B6, this one feature distinguishes Hole 1 from any of the other three holes.

If the run time system finds a hole and a B-type corner such that the corner is at B6's relative position and orientation, the hole is probably Hole 1 and the corner is probably B6. For a couple of reasons, the system cannot be absolutely positive that the pair of features it has found is B6 next to Hole 1. First, two hinges may be stacked in such a way that a B-type corner appears at the right distance and orientation with respect to another hole. Second, some noise along the boundary may look like a corner that just happens to be at the right distance and orientation with respect to a hole. Even though the system cannot be sure that the hole is Hole 1 and the corner is B6, finding a pair like that is generally sufficient evidence for hypothesising an occurrence of the object.

Having selected the B-type corner to distinguish Hole 1 from Hole 2, the next step to evaluate the contributions of that feature toward distinguishing other pairs and determining the object's orientation with respect to other holes. As already mentioned, the B6-type corner distinguishes Hole 1 from Hole 3 and Hole 1 from Hole 4. It can also be used to compute the orientation of the object relative to Hole 1. Since the other holes do not have matching B-type corners, this nearby feature does not help determine the orientation of the object with respect to them.

The first line in Fig. 36 summarises the contribution of this B-type corner near a hole. The goal of the selection procedure was to select a feature to distinguish Hole 1 from Hole 2 (written Disting. 1×2 in Fig. 36). The corner B6 relative to Hole 1 was selected. None of the other holes had matching B-type corners. Therefore, that feature distinguishes Hole 1 from Hole 2, Hole 1 from Hole 3, and Hole 1 from Hole 4. It also determines the orientation of the object with respect to Hole 1. The remainder of Fig. 36 summarises additional selections for hole-type focus features.

Since the B-type corner does not distinguish Hole 2 from Hole 3, the next subgoal of the feature selection procedure is to choose a feature that will do this. It chooses A1 relative to Hole 2 because it is the closest feature to either Hole 2 or Hole 3, whereas Hole 3 does not have a matching A-type corner. As indicated in Fig. 36

Goal	Selection	Additional Occurrences	Distinction						Orient.			
			1×2	1×3	1×4	2×3	2×4	3×4	1	2	3	4
Disting. 1×2	B6 wrt Hole1	—	*	*	*				*			
Disting. 2×3	A1 wrt Hole2	—	*			*	*			*		
Disting. 3×4	B1 wrt Hole4	—			*		*	*				*
Orientat. 3	Hole2 & Hole4 wrt Hole3	(H2 wrt H1 H1 & H2 wrt H2 H3 wrt H4)										
			3	2	2	1	3	2	2	2	1	2
ADDITIONS FOR A REDUNDANCY FACTOR OF 2												
Disting. 2×3	A3 wrt Hole3	A2 wrt Hole1	*		*	*		*	*	*	*	
			4	2	3	2	3	3	3	2	2	2
ADDITIONS FOR A REDUNDANCY FACTOR OF 3												
Disting. 1×3	B1 wrt Hole3	B2 wrt Hole4		*	*	*	*				*	*
Orientat. 2	A2 wrt Hole2	—	*		*	*				*		
			5	3	4	4	5	3	3	3	3	3

Fig. 36. Sequence of nearby features automatically selected for hole-type focus features.

none of the other holes have matching A-type corners. Corner A4 is at approximately the same distance from Hole 4 as A1 is from Hole 2, but its relative orientation is sufficiently different (as determined by the variances in the object model) to be distinguishable.

To differentiate between Hole 3 and Hole 4, the system selects B1 with respect to Hole 4. After this nearby feature has been added to the list and the contributions updated, the only subgoal left to be achieved is to establish the orientation of the object with respect to Hole 3. To satisfy this subgoal, the system selects the closest unused group of features near Hole 3. Since none of the features near Hole 3 have been used, the first group, which is a pair of holes, is added to the list. Hole 1 and Hole 4 each have one matching hole near them, while Hole 2 has two such holes. Therefore, locating these holes can distinguish Hole 1 from Hole 2, Hole 1 from Hole 3, Hole 2 from Hole 4, and Hole 3 from Hole 4. Since patterns of holes around all four focus features are rotationally asymmetric, they can be used to determine the orientation of the object with respect to their focus features.

This choice of holes near other holes points up a weakness of the current selection procedure. It is designed to use the closest unused feature that can achieve the goal. Otherwise, it could have selected A3 with respect to Hole 3 to determine the orientation of Hole 3. A3 is slightly farther from Hole 3 than the pair of holes. A3 is a better choice, however, because none of the other holes have matching A-type corners, which means that fewer features would be required at run time to perform the task.

Fig. 34 shows the clusters of features implied by the four types of nearby features that have been selected for a hole-type focus feature. These clusters are minimal, in the sense that missing one of the features at run time could lead to an ambiguity. A feature might not be detected for several reasons. It could be occluded by another part; the part could be defective in such a way that it does not have a B-type corner in the expected position; the feature might be out of the camera's field of view; it might be so distorted that the feature detector cannot recognise it. In the hinge example, missing B6 with respect to Hole 1 results in an ambiguity. The system expects to find B6 and Hole 2 near Hole 1. If B6 is not detected, the pair of holes is ambiguous; it could consist of Hole 1 and Hole 3, Hole 2 and Hole 3, or Hole 3 and Hole 4.

If the user of this automatic system wants to incorporate some redundancy into the locational process, he can request the system to include enough features to achieve each goal in two or more ways. The numbers under the columns in Fig. 36 indicate the number of discrete ways of achieving each subgoal. These numbers are referred to as redundancy factors. The first row of numbers summarises the contributions of the first four feature selections. There is at least one way to achieve each goal, two ways to achieve six of the goals, and three ways to achieve two of them. If the user wants at least two different ways to accomplish each goal, the system needs features to distinguish Hole 2 from Hole 3 and to determine the orientation of the object with respect to Hole 3. As indicated in Fig. 36 the system selects A3 with respect to Hole 3, which happens to satisfy both subgoals. To obtain a redundancy factor of 3, the system adds two more types of features.

The higher the redundancy factor, the lower the profitability of not detecting enough features to identify the focus feature. However, increasing the list of nearby features leads to larger graphs to be analysed, which in turn lengthens the processing time required to make hypotheses. In the case of the holes in the hinge part, the average graph sizes are 10.75, 11.75, and 13.00 nodes for redundancy factors of 1, 2, and 3, respectively. Sometimes the increase is more dramatic. For

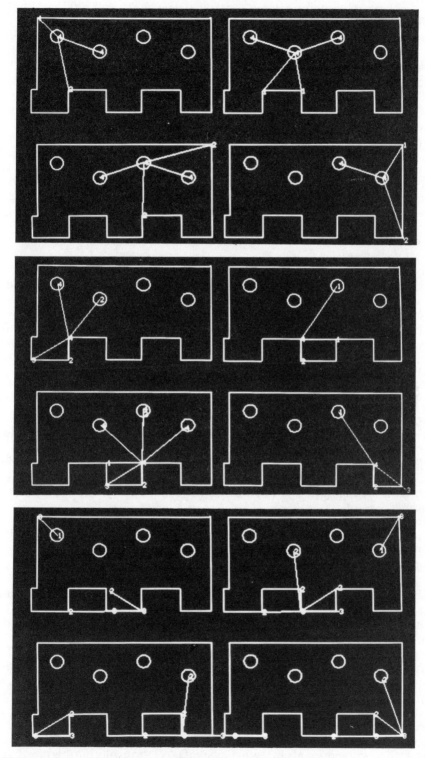

Fig. 37. Nearby feature clusters having a redundancy factor of three.

example, the graph sizes for A-type corners in the hinge are 5.50, 8.00 and 14.75 for redundancy factors of 1, 2, and 3.

The optimum redundnacy factor is a function of several variables, including the costs of building and analysing graphs to produce hypotheses, the costs of verifying hypotheses, and the probability of missing features at run time. For the current implementation, we have found that redundancy factors of 2 or 3 work well for images containing four or five overlapping objects, such as the hinge part. The cost of making a hypothesis is approximately five times the cost of verifying one, but the probability of not detecting a feature at run time is so high that the extra features are needed to ensure finding a sufficient set.

After selecting nearby features for a hole focus feature, the system selects nearby features for the other two potential focus features, A-type corners and B-type corners. Fig. 37 shows the clusters implied by the nearby feature selections for all three focus features. A redundancy factor of 3 was in these selections. These nearby features are the ones described in Fig. 9 and used for the examples in the run time section. In the next subsection we describe a procedure for deciding which focus feature to use first.

Focus feature ranking. The LFF training system ranks the foci according to the sizes of the graphs they imply, since the construction and analysis of graphs are the most time-consuming steps in the locational process. The system constructs the clusters of features expected around each occurrence of a focus feature (as shown in Fig. 37), and then computes the sizes of the graphs from the lists of interpretations for each feature. The final ranking of a focus feature is based on the average size of its implied graphs. For the clusters shown in Fig. 37 the average graph sizes are 13.00 nodes for holes, 14.25 for A-type corners, and 14.50 for B-type corners.

Discussion

The LFF system can easily be trained to locate partially visible two-dimensional objects. However, like all methods it is based on a set of assumptions that implicitly define the class of tasks it can perform. In this section we discuss these assumptions, their implications, and possible ways to eliminate them.

The basic assumptions of the current implementation of the LFF system are the following:

1. The objects rest on a plane in one of a few stable states.
2. The image plane of the camera is parallel to the plane supporting the objects.
3. The objects can be recognised as silhouettes in a binary image.
4. Each object contains several local features that are at fixed positions and orientations in the object's coordinate system.
5. The objects can be distinguished on the basis of relatively small clusters of nearby features.
6. Speed and reliability are important.
7. The more the training phase can be automated, the better.

The first and second assumptions combine to restrict the tasks to those that can be performed by two-dimensional analysis. The assumptions imply a simple one-to-one correspondence between features in an image and features on an object. There is a direct correspondence between distances and orientations in the image and distances and orientations on the support plane. The effects of a perspective projection are minimised. Since three-dimensional distances and orientations can be measured directly in range data, an LFF-type recognition strategy could be used to locate three-dimensional objects in range data. We are exploring this possibility.

The third assumption applies only to the current implementation. It is possible to implement an LFF system that uses grey-scale feature detectors. The types of object features and feature detectors would change, but the basic recognition strategy would remain the same. Such a system would be rendered particularly attactive by the inclusion of special-purpose hardware designed to locate corners and holes in grey-scale images.

The fourth assumption limits the class of the objects that can be recognised to those that have several local features, such as corners and holes. The LFF system cannot locate objects such as French curves that are characterised by large, continuous arcs. Other methods, such as generalised Hough techniques,[6] are better suited to this class of tasks.

The fourth assumption restricts the tasks to the recognition of rigid objects. We plan to explore ways to relax this constraint. In particular, we plan to investigate ways to locate objects with movable components. A simple strategy would be to ignore any features associated with such components. However, often these transient features contribute important, sometimes even crucial, constraints. How can they best be captured? The answer may lie in a multistage rcognition procedure. We have already implemented the LFF system as a two-step procedure in order to deal with objects that are not mirror-symmetric. We originally treated an object and its mirror as two separate objects. However, since the features and their relative distances are identical in both objects, we soon found that it was much more efficient to treat them as a single entity, relax our constraints on the relative orientations of features, and insert a second step in the recognition procedure that uses pairs or triplets of features to ascertain whether the object is right-side up or upside down.

The fifth assumption emphasises the fact that the LFF system is not designed to recognise an object by discerning one feature at one end of the part and another feature at the other end. It is designed to utilise local clusters of features. Figs. 38 to 41 illustrate this point. Fig. 38 shows two objects, one of which almost completely covers the second. The system recognises one of these objects but not the second, because the detectable features are too far apart; they are not part of a local cluster. This concentration on local clusters is not as restrictive as one might think. For, if the objects are only slightly farther out of alignment, as in Fig. 40, local clusters emerge and the system is able to locate both of them (see Fig. 41). The LFF system, like all systems for locating partially visible objects, is better suited to tasks in which the objects are mostly visible, as opposed to tasks in which objects are almost completely occluded.

The LFF system was designed to locate industrial parts that may contain several identical features or patterns of features. However, as stated in the fifth assumption, the efficiency of the system depends on the fact that the objects can be distinguished on the basis of small clusters. Large clusters take longer to find. A corollary to this statement is that the LFF system is slowed in its progress not by the number of objects to be recognised, but rather by the size of the ambiguous clusters. Fortunately, most parts – even machined parts that tend to contain patterns of identical features – have small distinguishing clusters.

The sixth assumption is somewhat of a catch-all, but the intent is to rule out highly parallel matching techniques that require massively parallel hardware to make them practical. Such devices and methods will certainly be developed, but, at least for the immediate future, we are limited to a sequential machine. In light of that fact, we want to take advantage of the opportunity available in many industrial tasks to analyse models of the objects to be recognised, so as to improve the speed and reliability of the recognition process.

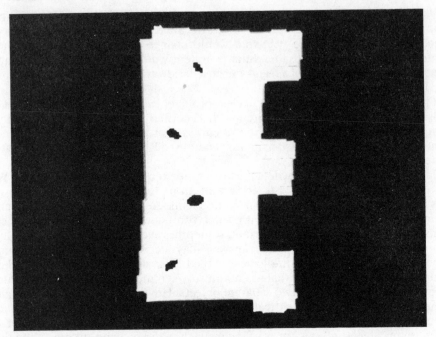

Fig. 38. Image of one hinge part almost directly on top of another.

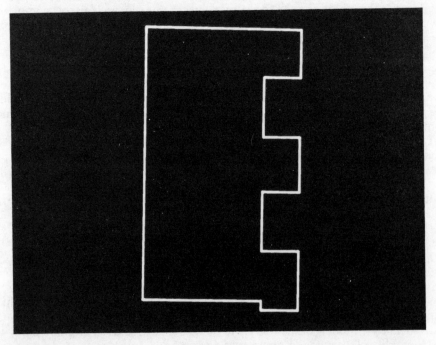

Fig. 39. Location of one of the hinge parts in Figure 38.

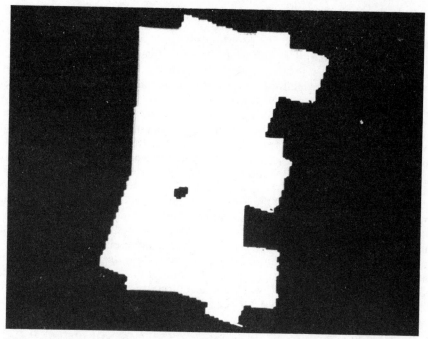

Fig. 40. Image of two hinge parts that are not quite as closely aligned as the pair in Figure 38.

Fig. 41. Locations of the two hinge parts in Figure 40.

We plan to continue the investigation of general-purpose techniques for analysing models of objects and selecting key features to be used at run time. We plan to extend the LFF selection procedure by having it explore more combinations of nearby features and by improving its evaluation of them. To this end we are considering techniques for incorporating the cost and reliability of locating features, as well as other techniques for rating features according to the number of goals they help achieve and how the accomplish this.

The seventh assumption is not really an assumption. It is rather a statement of our philosophy. We believe that industrial vision systems will gain wide acceptance only when they can be easily trained. The easier the training, the more tasks they will be used for. The concept of selecting key features automatically is not limited to LFF strategy. A similar selection process could be used to select the best pairs of local features for a histogram-type matching scheme. Since the number of feature pairs increases as the square of the number of features, a judicious reduction in the number of features could generate a substantial saving.

In conclusion, it should be emphasised that the LFF system is more than just an efficient technique for recognising and locating a large class of partially visible objects. It is also a semi-automatic programming system for selecting key features to be used in its own locational strategy. From a global perspective, however, it represents but one small step toward the development of fully automatic systems with the dual capability of both strategy and feature selection.

APPENDIX
Maximal-clique algorithm

For completeness we describe an algorithm for locating all maximal cliques (i.e., completely connected subgraphs) of a graph. The algorithm is essentially a restatement of one described by Johnston.[23] A more detailed description of this algorithm, its derivation, and uses can be found in[10].

The algorithm is a recursive procedure of three parameters: C, P, and PuS. All three are sets that can be efficiently represented as arrays of bit strings. The set Q represents a clique that is to be extended into a maximal clique. The set P contains all the 'prospects' for extending the clique C. That is, each member of P is directly connected to all nodes in C. The members of P may not be connected to each other, but all of them are directly connected to all nodes in C. The set PuS is a union of P and the set containing all 'suspects' that could have been used to extend C, but which were arbitrarily left out. The algorithm is as follows:

```
      RECURSIVE PROCEDURE
maxCliques (set C,P,PuS);
      BEGIN 'maxCliques'
      setElement X,Y;
      IF emptySet(PuS) THEN List-C-as-a-Maximal-Clique
          ELSE BEGIN 'chooseElement'
              X _ chooseOneElementFrom(PuS);
          ForEach Y IN intersect(P, complement(neighbours(X))) DO
                  BEGIN 'recursiveCall'
                  P _ deleteElementFromSet(Y, P);
                  maxCliques(add ElementToSet(Y, C),
                              intersect(P, neighbours(Y)),
                              intersect(PuS, neighbours(Y)));
                  END 'recursiveCall';
              END 'chooseElement';
      END 'maxCliques';
```

The procedures designated by long names perform the standard operations on sets, while the neighbours procedure returns the set of nodes that are directly connected to the node represented by the argument. To list all maximal cliques in a graph, set up the global data structure that represents the graph and make the call MaxCliques(emptySet, setOfAllNodes, setOfAllNodes).

It is tempting to try to increase the efficiency of the algortihm by carefully choosing X (e.g., choosing the member of PuS that has the smallest number of neighbours). However, the time required to select X invariably exceeds the time saved by not pursuing some dead-end paths,. The simplicity of the algorithm appears to be the key to its speed. Johnston[23] reported that it only took 0.569 seconds to produce and count all 687 maximal cliques in a graph that contained 48 nodes and in which 50 percent of the pairwise compatibilities held. His program was written on an ICL 1900-series computer that has 48-bit double words and a memory reference time of 300 nanoseconds.

If procedure calls are expensive, the efficiency of this algorithm can be improved by unwrapping some of the recursion. For example, if the parameter PuS in the recursive call is empty, the clique can be listed as a maximal clique without making the call. This type of call can be avoided by moving the test from the beginning of the algorithm to just before the recursive call. If the argument PuS is not empty, but the argument P is, the call would not produce anything. Such calls can also be avoided by inserting a test before the call.

If the only objective of the calling procedure is to find either the largest maximal clique or those containing some minimum number of nodes, the efficiency of the algorithm can be improved by stopping the recursive building of cliques when the number of nodes in C plus the number of nodes in P is less than the appropriate number. Here too, if this test is placed before the recursive call, unnecessary calls can be avoided.

References

[1] P. Kinnucan, 'How Smart Robots are Becoming Smarter', *High Technology*, p. 35 (September 1981).

[2] Bausch and Lomb Omnicon Pattern Analysis System, Analytic Systems Division Brochure, 820 Linden Avenue, Rochester, New York (1976).

[3] A. G. Reinhold and G. J. VanderBrug, 'Robot Vision for Industry: The Autovision System', *Robotics Age*, pp. 22-28 (Fall 1980).

[4] R. M. Karp, 'Reducibility Among Combinatorial Problems', *Complexity of Computer Computations*, pp. 85-103 (Plenum Press 1972).

[5] W. A. Perkins, 'A Model-Based Vision System for Industrial Parts', *IEEE Transactions on Computers*, Vol. C-27, pp. 126-143 (February 1978).

[6] D. H. Ballard, 'Generalizing the Hough Transform to Detect Arbitrary Shapes', *Pattern Recognition 13*, Vol. 2, pp. 111-112 (1981).

[7] S. Tsuji and A. Nakamura, 'Recognition of an Object in a Stack of Industrial Parts', *Proceedings of IJCAI-75*, Tbilisi, Georgia, USSR, pp. 811-818 (August 1975).

[8] S. W. Holland, 'A Programmable Computer Vision System Based on Spatial Relationships', General Motors Research Publication GMR-2078 (February 1976).

[9] A. P. Ambler, et al, 'A Versatile Computer-Controlled Assembly System', *Proceedings of IJCAI-73*, Stanford, California, pp. 298-307 (August 1973).

[10] R. C. Bolles, 'Robust Feature Matching Through Maximal Cliques', *Proceedings of SPIE's Technical Symposium on Imaging Applications for Automated Industrial Inspection and Assembly*, Washington, D.C. (April 1979).

[11] S. W. Zucker and R. A. Hummel. 'Toward a Low-Level Description of Dot Clusters: Labeling Edge, Interior, and Noise Points', *Computer Graphics and Image Processing*, Vol. 9, No. 3, pp. 213-233 (March 1979).

[12] S. T. Barnard and W. B. Thompson, 'Disparity Analysis of Images', *IEEE Transactions on Pattern Analysis and Machine Intelligence* Vol. PAMI-2, No. 4, pp. 333-340 (July 1980).

[13] R. O. Duda and P. E. Hart, 'Use of the Hough Transform to Detect Lines and Curves in Pictures', *Communications of the ACM*, Vol. 15, No. 1, pp. 11-15 (January 1972).

[14] S. Tsuji and F. Matsumoto, 'Detection of Ellipses by a Modified Hough Transformation'. *IEEE Transactions on Computers*, Vol. C-27, No. 8, pp. 777-781 (August 1978).

[15] G. Gleason and G. J. Agin, 'A Modular Vision System for Sensor-Controlled Manipulation and Inspection', *Proceedings of the 9th International Symposium on Industrial Robots*, Washington, D.C., pp. 57-70 (March 1979).

[16] J. Hill, 'Dimensional Measurements from Quantized Images', in 'Machine Intelligence Research Applied to Industrial Automation', Tenth Report, NSF Grant DAR78-27128, pp. 75-105, SRI International, Menlo Park, California (November 1980).

[17] T. G. Evans, 'A Program for the Solution of a Class of Geometric-Analogy Intelligence-Test Questions', *Semantic Information Processing*, pp. 271-353, (MIT Press, Cambridge, Massachusetts, 1968).

[18] J. Gips, 'A Syntax-Directed Program that Performs a Three-dimensional Perceptual Task', *Pattern Recognition*, Vol. 6, pp. 189-199 (1974).

[19] T. Kanade, 'Recovery of the Three-dimensional Shape of an Object from a Single View', *Artificial Intelligence*, Vol. 17, pp. 409-461 (1981).

[20] H. Wechsler, 'A Structural Approach to Shape Analysis Using Mirroring Axes', *Computer Graphics and Image Processing*, Vol. 9, No. 3, pp. 246-266 (March 1979).

[21] R. Silva, 'The Detection of Rotational Object Symmetries from Multiple Object Views', in the Seventh Annual Report on General Methods to Enable Robots with Vision to Acquire, Orient, and Transport Workpieces by J. Birk et al, University of Rhode Island (December 1981).

[22] R. C. Bolles, 'Local Visual Features', in the SRI Ninth Report on Machine Intelligence Research Applied to Industrial Automation, pp. 77-108 (August 1979).

[23] H. C. Johnston, 'Cliques of a Graph – Large or Small', Draft, Queen's University of Belfast (1975).

THE COUPLING OF A WORKPIECE RECOGNITION SYSTEM WITH AN INDUSTRIAL ROBOT

G. Nehr and P. Martini, Institut für Informatik, University of Karlsruhe, Germany

Introduction

When manufacturing medium and small lot sizes, workpieces can automatically be produced by CNC machine tools. However, the problem of automatic conveying of workpieces to and from the machine tools has not been solved to date. This task can only be achieved by the application of flexible, computer controlled material handling systems, which are easily adaptable to different workpieces.

Flexibility of these systems, however, can only be guaranteed, if they are equipped with sensors by which the workpieces can be recognised and their position can be determined. For this purpose, optical sensors are suitable because they can be universally applied. This paper describes the architecture of a sensor-controlled industrial robot and deals especially with the problems which will arise when robots are connected to sensor systems.

The recognition system

The architecture of the recognition system is described in the block diagram Fig. 1. It consists of an optical sensor which is coupled to a microcomputer via special hardware processors. A 4096 photodiode array is used as an optical sensor which is located in a camera housing together with the scanning module and the signal amplifier.

Since the sensor consists of a 64 × 64 square matrix and because of the predefined scanning mode of the sensor signals the picture field can be described in cartesian coordinates (Fig. 2). The picture, which is scanned by the camera, is first digitised with the help of an adaptive threshold. Then its binary information is stored for further processing in a memory.

The recognition procedure is subdivided into two parts, a teach-in phase and a processing phase. During the teach-in phase the system is taught the identity of the object by showing the workpiece to the camera. A set of features M is derived from this pattern and is stored in the object memory section of the computer as reference.

The feature vector is obtained by calculating the object's surfaces with the help of circles which are concentrically located about the object's centre of gravity (Fig. 3).

This research work was performed at the Institut für Informatik III under Prof. Dr.-Ing. U. Rembold at the University of Karlsruhe, and was funded by the BMFT/PDV.

In order to calculate the orientation of the object within the picture field another set of features $C_R(n)$ is produced by a special scanning technique applied to the picture (Fig. 4).[1]

During the work phase a set of features M_A is calculated on the basis of the presented, unknown pattern and is compared with all other sets of features M_1, M_2 . . . , M_n stored in the object memory.

This comparison supplies criteria for the decision whether the unknown pattern can be referred to an object class of the set M_1, M_2, . . . , M_n or not. After the pattern is identified, the orientation of the object in comparison to the reference position obtained from the teach-in phase is calculated. As in the teach-in phase, the term $C(k)$ is determined, which is, except for a phase shift, identical to the term $(C_R(k)$, stored during the teach-in phase (Fig. 5).

This phase shift is the angular orientation of the object compared to its reference position.[2]

After completion of the recognition all features are present which are necessary to control grasping in a handling system. They are:

○ object class,
○ object position,
○ object orientation.

The robot system

The handling system coupled to the sensor was a Siemens industrial robot. The mechanical design had four degrees of freedom: a vertical lift axis, a swivel axis, a horizontal transfer axis and a gripper rotation axis.[3] The gripper was equipped with suction cups allowing it to grasp flat workpieces (Fig. 6). All axes of the equipment were operated by DC pancake motors. The operating range of each axis is approximately + 32000 displacement units in regard to a mechanical reference point.

A Siemens minicomputer PR 310 was used as the heart of the control unit.

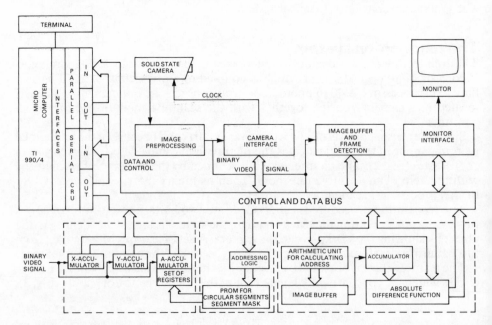

Fig. 1. Architecture of the recognition system.

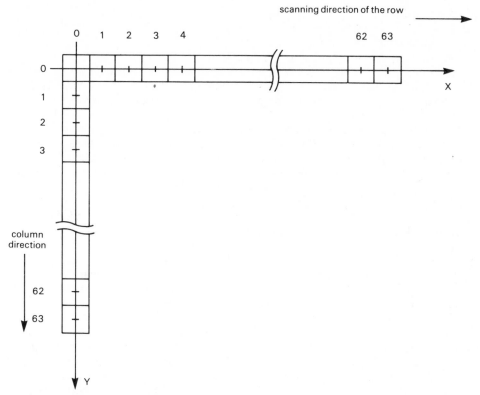

Fig. 2. Cartesian coordinate system of the sensor.

Beside a central memory of 8k words, the hardware also contained interfaces for the robot's four axes and the input-output modules. The input-output modules were used to transfer the control signals between the control unit and the operator panel and the robot and the peripherals respectively (Fig. 7).

The function of the control unit can be divided into the following sub-functions:
- ○ processing of program information,
- ○ processing of manual-entered information,
- ○ action processing,
- ○ processing of control signals,
- ○ control of the operator panel,
- ○ coordination of the above mentioned sub-functions.

The majority of these sub-functions are realised in software modules and contained in a program system (Fig. 8).

When starting the equipment or if an error occurs during the execution of the program system, the highest priority level '0' will become active. The cyclical part of the program is started every 5ms by a clock signal run at priority level 2. For control of the robot axes, interaction between the clock cycle signal and the beginning of the program is necessary. Therefore all software modules are designed such that their execution is shorter than 5ms.

Interfacing of the sensor and the industrial robot

It is necessary to observe three aspects when coupling a sensor with an industrial robot:

○ physical interface,
○ data transfer protocol,
○ coordinate transformation.

Because the development stage of industrial robots with optical sensors is only just beginning, to date there has been no standardisation of the problems mentioned above. The designer of such a system therefore relies on his own intuition. However, he is limited by hardware and software support of the sensor system as well as of the control unit of the manipulator.

When the Siemens industrial robot was interfaced with the sensor system, it had to be taken into account that the robot control unit should be operated within the frame of the 5ms-clock cycle. To keep within this time the coordinate transformation had to be carried out by the sensor computer. Data transfer between the sensor system and the control unit was carried out by a handshake procedure, but during each 5ms cycle, only one byte could be transferred.

For the sensor-robot system the following concept was developed. For each workpiece known to the sensor system, an action program was assigned by means of a program number. The grasping position of the workpiece, which does not correspond with the centre of gravity, is calculated. This is done by bringing the gripper under hand control to the grasping position and by transferring the coordinates of this position to the sensor computer. From these coordinate axes the grasping position is calculated in reference to the sensor system. With this information the positional values are determined, which are needed for the calculation of the grasping position in the work phase.

During the work phase, the control unit is waiting for the transfer of the grasp coordinates. After a workpiece is identified by the sensor, first the grasping position has to be calculated in reference to the sensor system, and then this grasping position is transferred to the robot's coordinate system. The coordinate axes obtained by this transformation, are transferred to the control unit together with the program number of the action program which was assigned to the workpiece in the teach-in phase. The action program, modified by the new coordinate axes, can then be processed by the robot control.

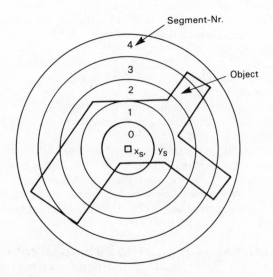

Fig. 3. Calculating the feature vector M.

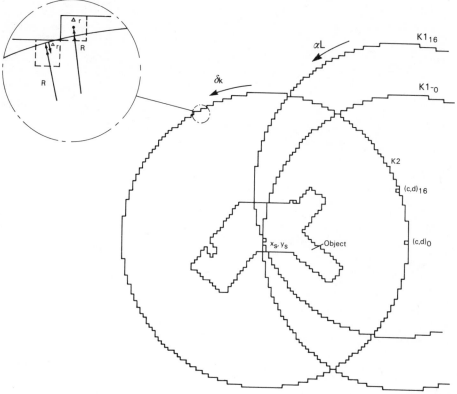

Fig. 4. Calculating the feature vector $C_R(k)$

Coordinate transformation during the teach-in phase

The teach-in phase of the coupled sensor-system used the following procedure:
○ teaching the workpiece to the sensor system,
○ determining the grasp position with the help of the robot,
○ planning an action program for handling of the workpiece.

In the following paragraphs only the calculation of a grasp position is described. To do this we start with a setting of the robot/vision system shown in Fig. 9. The location of the corresponding cartesian sensor coordinate system with regard to the robot coordinate system is shown in Fig. 10. The sensor coordinates are described by X and Y, the robot coordinates by T and S. In the robot system the grasp position G is described by the position of the transfer axis T_G and the position of the swivel axis S_G when the gripper is above the grasp position. The distance H between G and M can be calculated with regard to the robot system:

$$H_R = T_G + T_O, \text{ increment units of the transfer axis } (I_T)$$

in reference to the sensor system:

$$H_S = \frac{H_R}{F_T}$$

The constant value T_O describes the distance between the gripper and the centre

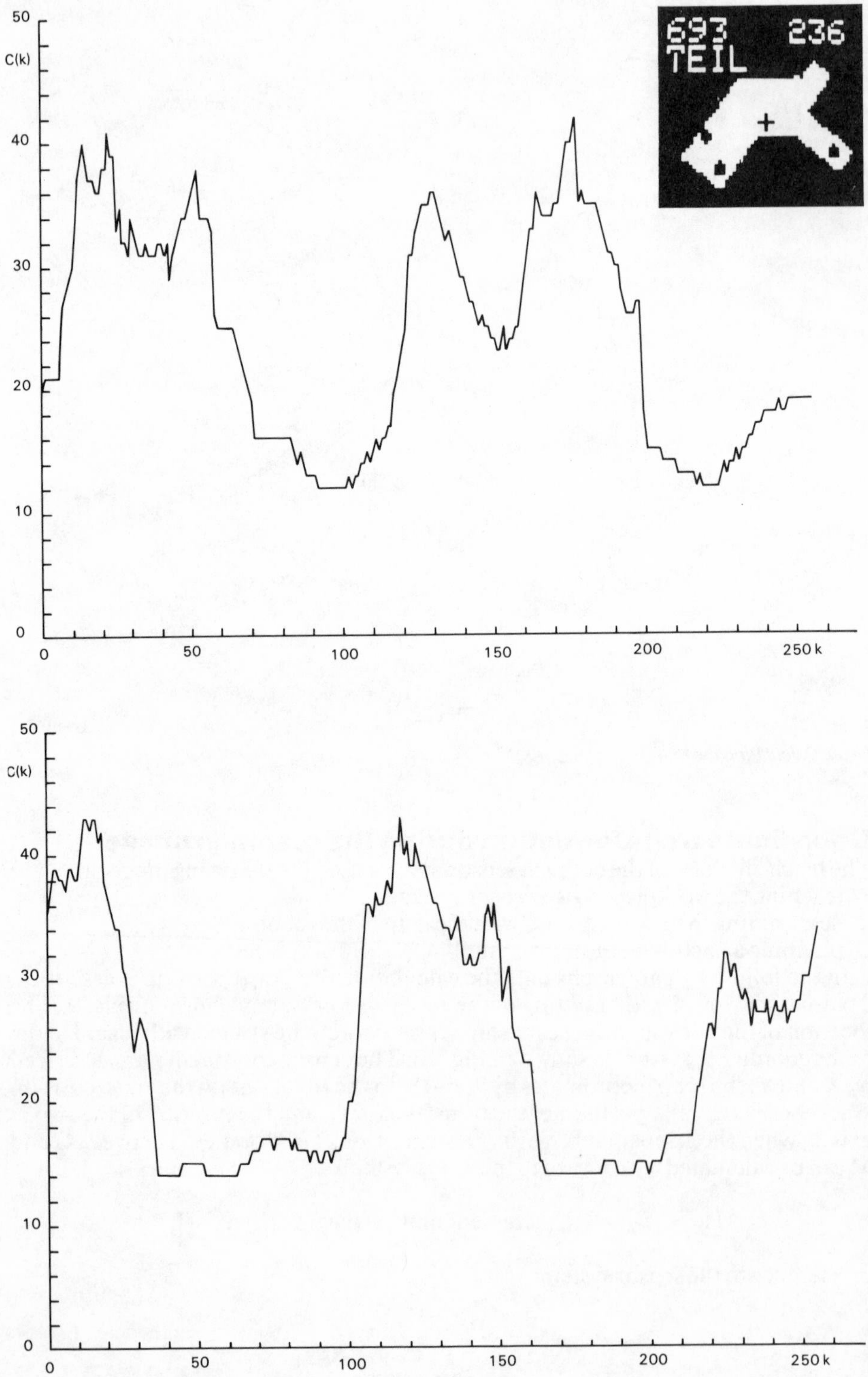

Fig. 5. Polar coding of an object (a) in the reference position; (b) during the work phase.

of rotation of the swivel axis, when the axes are at the reference point. It is measured in increment units of the transfer axis. The constant value F_T represents the ratio of the transfer increment units described by the length of one pixel ($I_T/BPKT$).

The following equation describes the angle:

$$\alpha = \frac{S_O - S_G}{F_S} \quad \text{(degrees)}$$

The constant value S_O describes the position of the swivel axis when it is parallel to the sensor-X-axis. The constant value F_S shows the number of increments of the swivel axis per degree of angle ($I_S/\text{deg.}$). The coordinates of the point G in regard to the sensor-coordinate system can be calculated from:

$$\triangle X = H_S \cdot \cos(\alpha) \quad \text{(BPKT)}$$
$$\triangle Y = H_S \cdot \sin(\alpha) \quad \text{(BPKT)}$$

Coordinate transformation during the operating phase of the sensor

During the operating phase the following tasks are carried out:
○ calculation of the centre of gravity P_A (X_P, Y_P),
○ identification of the workpiece,
○ calculation of the rotational position of the workpiece,
○ calculation of the grasp position with regard to the sensor G_A (X_G, Y_G),
○ transformation of the coordinates of the grasp position to the coordinates of the robot system.

The values obtained during the operating phase are marked by the subscript A. Fig. 11 shows the interrelation of the different coordinates and the important system parameters during the operating phase. After identification of the workpiece and after calculation of its rotational angle δ with regard to the reference position, the grasp position of the workpiece with regard to the sensor position has to be calculated.

For the workpiece orientation γ_A one obtains:

$$\gamma_A = \gamma_L + \delta$$

in which γ_L is the workpiece orientation and δ the rotational angle of the workpiece, both in regard to the reference position.

The grasp position G_A (X_G, Y_G) is then calculated from:

$$\triangle X = D \cdot \cos(\gamma_A)$$
$$\triangle Y = D \cdot \sin(\gamma_A)$$
$$X_G = X_P + \triangle X$$
$$Y_G = Y_P + \triangle Y$$

with (X_P, Y_P) denoting the centre of gravity of the object obtained during the operating phase and D the distance between the centre of gravity and the grasp position.

Now the grasp position has to be transferred to the robot coordinate system. For this the following calculations are made:

☐ Position of the swivel axis,

$$S_A = S_O + FS \quad \arctan \frac{Y_G - Y_O}{X_O - X_G} \tag{I$_S$}$$

Thus the following results are obtained:

$$X_G = X_O - \Delta X$$
$$Y_G = Y_O - \Delta Y$$

in which the coordinates X_O, Y_O describe the centre of rotation of the swivel axis in regard to the sensor system. Thus the grasp position of the reference position of the workpiece is known. During the work phase the different position of each workpiece has to be calculated on the basis of the reference position. For this task the

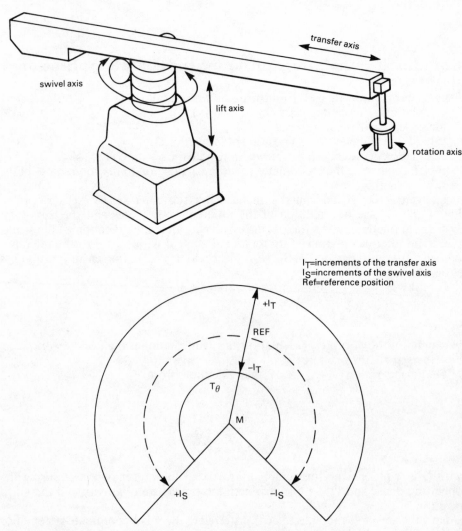

I_T=increments of the transfer axis
I_S=increments of the swivel axis
Ref=reference position

Fig. 6a. Mechanical components of the robot. 6b. The coordinate system of the robot.

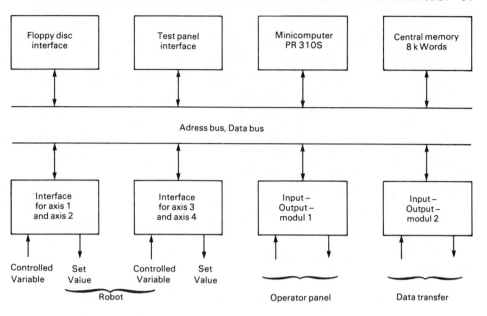

Fig. 7. Components of the robot's control unit.

distance D between the grasping position G (X_G, Y_G) and the centre of gravity P (X_P, Y_P) and also the workpiece orientation are calculated during the teach-in phase. The distance D is obtained from:

$$D^2 = (X_P - X_G)^2 + (Y_P - Y_G)^2 \qquad \text{(BPKT)}$$

and the workpiece orientation is:

$$\gamma = \arctan \frac{Y_P - Y_G}{X_G - X_P} \qquad \text{(degrees)}$$

For the calculation of the grasp position and for its transformation to the robot coordinate system the following additional values have to be stored:

○ distance D between the grasp position and the centre of gravity   (BPKT),
○ orientation of the workpiece γ_L (degrees),
○ angle (α_L) (degrees),
○ number of the action program N,
○ position of the vertical lifting axis H_L (I_H),
○ position of the rotation axis R_L (I_R).

Here the index L describes the values which are obtained during the teach-in phase.

□ Position of the transfer axis:

$$T_A = F_T \cdot \left[(X_0 - X_G)^2 + (Y_0 - Y_G)^2 \right]^{\frac{1}{2}} - T_0 \qquad (I_T)$$

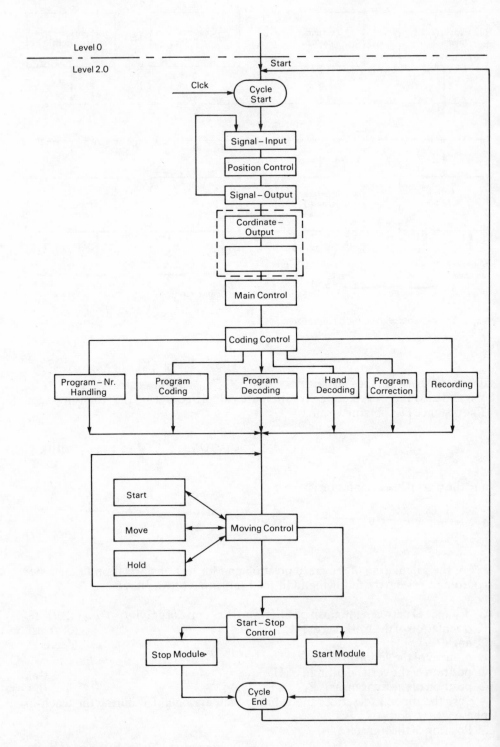

Fig. 8. Block diagram of the control software.

☐ Position of the rotational axis:

$$R_A = R_L + F_R \cdot (\delta + \alpha_A - \alpha_L) \qquad (I_R)$$

$$\alpha_A = \frac{S_A - S_O}{FS} \qquad \text{(deg.)}$$

☐ Position of the vertical lift axis:

$$H_A = H_L$$

and the number of the action program is:

$N_A = N_L$,
R_L is the position of rotational axis obtained from the teach-in phase,
F_R number of rotational increments per degree of angle. (I_R/degree),
H_L position of the lift axis obtained from the teach-in phase.
    The values S_A, T_A, H_A, R_A and N_A are then passed over to the robot control unit. Now the robot is able to move to the calculated grasp position.

Measuring the robot position in regard to the coordinate system of the sensor

In the two previous sections, constant values which have to be measured or calculated beforehand, had been used for the calculation and transformation of the grasp position. These constant values can be defined as:
(a) robot constant values, and
(b) constant values of the interfaced sensor-robot-system.

(a) T_O distance between the centre of the gripper in the reference position of the robot and the centre of rotation of the swivel axis. T_O is measured in increments the transfer axis units (I_T).

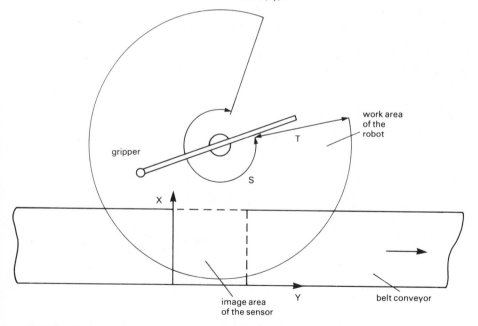

Fig. 9. Geometric set-up of the robot/vision system.

F_S number of swivel axis increment units per degree of angle (I_S/deg.).

F_R number of rotational increment units of the gripper per degree of angle (I_R/deg.).

(b) F_T number of horizontal transfer axis increment units per pixel units (I_T/BPKT).

S_O the angular offset of the swivel axis when it is located parallel to the sensor X-axis.

X_O, Y_O coordinates of the centre of rotation of the swivel axis in reference to the sensor system.

The constant values mentioned under (a) were measured directly at the robot. In order to obtain T_O, all axes had to be moved to the reference point. Thereafter the distance between the centre of the gripper and the centre of rotation of the swivel axis was measured and multiplied by the number of increments of measurement units. The constant F_S and F_R were obtained as follows:

The robot was brought to a reference position. At this position the value of the controlled variable of the swivel control loop was determined. Then the arm was rotated 180° and the new controlled variable was read. Thereafter the difference was taken between these two values and divided by 180. The same procedure was used to determine F_R for the rotation axis.

The second group constant values had to be obtained from the sensor-robot-system. For that purpose both the sensor and the robot were brought to a fixed position. With the help of the robot, a disc representing an object was moved between the three points P_1, P_2 and P_3, all of which were in the field of vision of the camera (Fig. 12). A disc was chosen because it has the following criteria:

○ the possibility of an error occuring when calculating the centre of gravity is small,

○ there are no geometric changes when rotated,

○ the grasp position can easily be defined as the centre of gravity of the disc.

The coordinate values (S_1, T_1), (S_2, T_2) and (S_3, T_3) are determined for each point. A prerequisite for this calibration procedure is that the position of the robot between P_1 and P_2 can only be altered by moving the transfer axis. In addition between the points P_2 and P_3 the robot is only allowed to move about its swivel axis. Because both the centre of gravity and the grasp position are identical with the disc, the grasp position with regard to the sensor system can be obtained by calculating

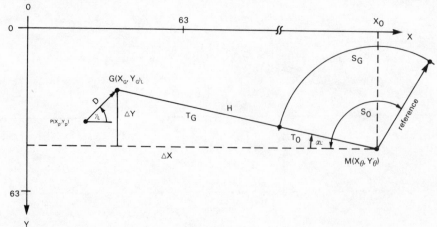

Fig. 10. Location of the sensor coordinate system in regard to the coordinate system of the robot.

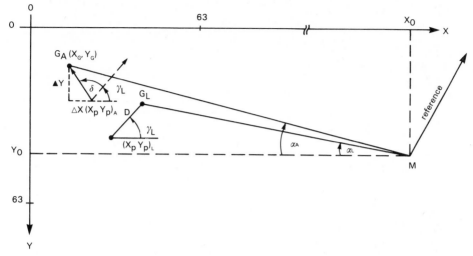

Fig. 11. Coordinate transformation during the work phase.

the centre of gravity at the points P_1 to P_3. Thus a fixed assignment between sensor and robot coordinates is obtained.

The constant values can be calculated as follows:

$$F_T = \frac{T_1 - T_2}{[(X_1 - X_2)^2 + (Y_1 - Y_2)^2]^{\frac{1}{2}}} \qquad (I_T/\text{BPKT})$$

In this equation the denominator is the equation of the circle:

$$(X_O - X_2)^2 + (Y_O - Y_2)^2 = r^2$$

and

$$(X_O - X_3)^2 + (Y_O - Y_3)^2 = r^2$$

with

$$r = \frac{T_2 + T_O}{FT}$$

Thus the centre of rotation of the swivel axis X_O, Y_O with regard to the vision system can be calculated.

Conclusion

The operation of the robot-sensor-system was demonstrated in connection with a belt conveyor. Different parts randomly oriented were moved on a conveyor. The vision system recognised the parts, their location and orientation. It advised the robot to pick up the parts and to place them in a defined position on a work bench.

This task hitherto could only have been accomplished by tools especially adapted to the workpiece. In general such tools are only economical when used for production runs with high piece rates. With the help of a sensor controlled robot, handling of a broad spectrum of workpieces may be possible, eliminating the necessity for special tools, so that the system can be used for low medium volume production.

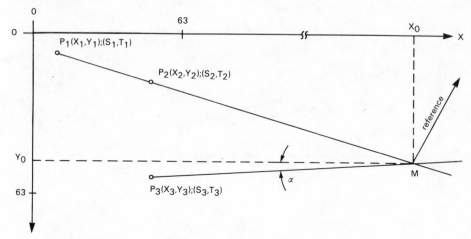

Fig. 12. Calibration of sensor and robot coordinate system.

References

[1] K. Armbruster, P. Martini, G. Nehr, U. Rembold and W. Olzmann, 'A very fast vision system for recognising ports and their locations and orientation', *Proc. 9th Int. Symp. Industrial Robots*, pp. 265-280, Washington, DC, USA, SME, March 1979.

[2] G. Nehr and P. Martini, 'Recognition of angular orientation of objects with the help of optical sensors', *The Industrial Robot*, Vol. 6, No. 2, pp. 62-69 (June 1979).

[3] K. Feldman, *et. al.* 'Einsatz des Siemens-Industrieroboters zur Maschinenbeschickung', *Proc. 8th Int. Symp. Industrial Robots*, pp. 841-849, Stuttgart, Germany, IFS (Publications) (May 1978).

[4] H. Becker and G. Leusch, 'Freiprogrammierbare Steuerung für Industrieroboter', *Siemens-Zeitschrift* Vol. 51 No. 6, pp. 463 (1977).

EVALUATING VISION SYSTEM PERFORMANCE

C. A. Rosen and G. J. Gleason, Machine Intelligence Corporation, USA

Reprinted by courtesy of the Society of Manufacturing Engineers, Dearborn, Michigan, USA from SME Technical Paper MS 80-700.

Abstract

A simple method is proposed for comparing the relative performance of machine vision systems designed for programmable automation. Some major performance criteria are the degree of discrimination between patterns and the execution time required to determine the identity, the state, and the position of workpieces.

The method is based on the use of a small set of two-dimensional test figures with readily-reproducible silhouettes. Workpieces with multiple internal regions of interest are simulated by a circular disc with multiple holes, increasing complexity represented by an increase in the number of holes, with a consequent increase in the required processing time for analysis.

The method of measurement is described, the geometric test shapes are illustrated, and typical performance measurements are reported for a commercial machine vision system.

Introduction

Machine vision, based on the processing and interpretation of electro-optical (television) images is being introduced into practice for a number of diverse programmable automation applications,[1-5] these include inspection, material handling, and robot assembly tasks. The first generation of available vision machines are designed to recognise and identify workpieces, their stable states, and to determine their positions and orientations under the following major constraints:

☐ Each workpiece is supported in one of a small number of stable states. (A work-piece hanging on a hook, able to turn, sway or swing has an infinite number of states and does not satisfy this constraint).

☐ A binary image, representing the silhouette of the workpiece can be extracted by enhancing the optical contrast between workpiece and background by suitable lighting techniques.

☐ Each workpiece is separated (not touching) from every other workpiece in the field of view. No overlapping of workpieces is permitted.

The above constraints define the conditions for present cost-effective vision systems dealing with binary images. The complexity of a visual scene under the above conditions depends on the number of separate entities (or 'blobs') that must be analysed in a given scene, and the number of states that must be discriminated for each workpiece. One workpiece may have several internal regions of interest, such as holes, for which some measured features are important for both recognition and inspection. Further, to provide visual sensory information for an industrial robot, each stable state of the workpiece must be recognised, and position and orientation determined.

A semi-quantitative classification of scene complexity for this type of simple vision has been described by Foith, *et. al.*[6] in terms of simple geometric relations between workpieces in the field of view. These range from the simplest case of one randomly-positioned workpiece in the field of view, to the most complex – in which several workpieces are either touching or overlapping each other. They distinguish one intermediate case of wide practical importance, namely, when several non-contacting non-overlapping workpieces are in the field of view and cannot be separated by non-overlapping rectangular bounding 'windows'. Two cases of such scenes are shown in Fig. 1. In Fig. 1(a) two workpieces are shown (non-touching) but intertwined such that the dotted rectangular 'windows' overlap. In Fig. 1(b) a disc with two holes represents three discrete regions of interest, (three 'blobs') – the disc and the two holes. These images can be analysed using connectivity analysis[3, 7] which automatically groups together picture points to form the regions of interest, that we have labelled 'blobs'. It is plausible to assume that one measure of complexity is the number of 'blobs' that must be processed to analyse the scene. A second measure is the number and type of geometric features that must be extracted to characterise or discriminate each state. Such features may include area, perimeter, second moments, maximum and minimum radii from centroid, number of internal holes (or 'blobs'), and others. A third measure of complexity is the amount of total data that must be processed per scene, that is, the total number of bits, whether binary, grey scale, colour or range, that describe the object of interest as distinguished from its background.

It is difficult to define an analytical expression to quantify the complexity of the rich variety of scenes that are possible. We propose, instead, a standard reproducible set of two-dimensional test figures which can be used to obtain

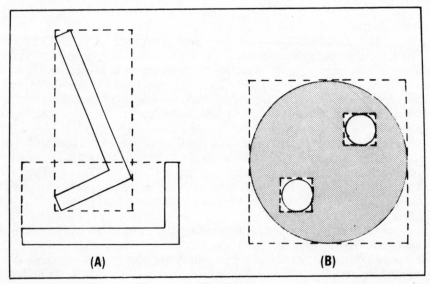

Fig. 1. Complex objects requiring connectivity or 'blob' analysis.

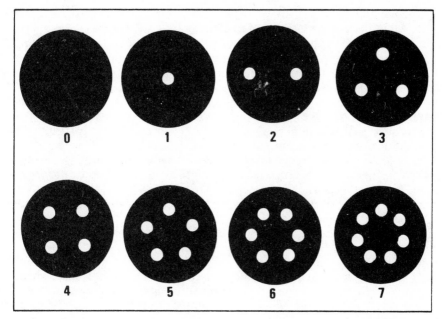

Fig. 2. Set of discs with variable number of holes.

All holes are $\frac{1}{4}$ inch diameter.
The centres of all holes lie on a
1 inch diameter circle.

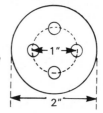

comparative performance data of different vision systems, such data related to the three complexity measures described above. Undoubtedly, other measures will be proposed and added as we learn more about appropriate classification of imagery and can better define what we mean by complexity.

Proposed method of test and pattern sets

It is proposed to use two sets of two-dimensional black and white geometric figures as test patterns for evaluating a few major performance factors of vision systems. The first set, shown in Fig. 2, is designed to measure execution time for recognition and measurement of important parameters as a function of complexity due to increase in the number of regions of interest ('blobs'), while total binary bits processed and number of discrimination features remain constant. The first set is composed of eight 2in. diameter discs, each with a different number of $\frac{1}{4}$in. diameter holes, ranging from zero to seven holes. The discs can be readily reproduced with high accuracy, photographically, providing high contrast binary patterns. They are of constant total area, therefore, requiring the processing of a constant number of binary bits. There is little change in the general shape of the different patterns in the set, such that the discrimination between each pattern will not be overly sensitive to selection of discrimination features.

The second set, shown in Fig. 3, are designed to present four subsets of patterns with significantly different shapes to test the efficiency of the set of discriminating features selected for differentiation. There are four different shapes, in subsets of two: There are two circular discs, two squares, two rectangles, and two equilateral triangles. All patterns have a constant total area of π square inches. Each subset of two of a kind have one pattern with no holes, and the other with one $\frac{1}{4}$in. diameter hole. Thus, this test should yield some insight into the relative worth of a selected set of features while the area (number of bits processed) remains constant, and the number of 'blobs' are two or less.

Experimental results

The performance of a commercial vision system[8] using these two sets of test patterns is illustrated in Fig. 4 and Table I.

In Fig. 4 the system was trained to adequately discriminate randomly-positioned discs with varying number of holes, using a nearest-neighbour classification.[7] A set of eleven features (Fig. 4) for one series of measurements was used. A single distinguishing feature, the total area, was used in another series of measurements to provide a basis for comparison. The 2in. diameter discs were randomly-positioned in a 4in. × 4in. field of view, imaged by a 128 × 128 element solid state camera (GE TN2200).

Curve 'A' indicates recognition times increasing from approximately 700 milliseconds for the disc with no holes to 850 milliseconds for the 7-hole disc, when the eleven features are used. Because the identification is based on eleven measurements, the discs are identified with high reliability. Further, complete data as to position and orientation of each important region (the whole disc and each hole) is available at the output terminals through the use of an appropriate communication protocol.

Curve 'B', the reference curve, shows recognition time of an approximately constant 250 milliseconds for all samples. This test used only the net area feature,

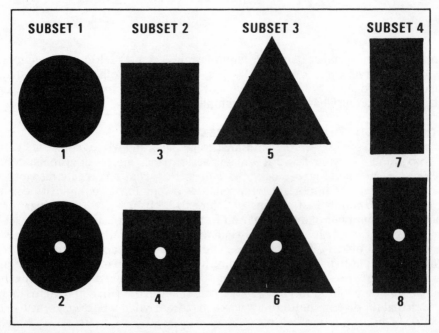

Fig. 3. Subsets of patterns with equal area, varying shapes.
Notes: (1) The area of each pattern is π square inches.
      (2) Holes are $\frac{1}{4}$ inch diameter.

i.e., it counted the number of bits representing the disc area, without the hole area, thus permitting rather poor discrimination based on small differences of net area.

Table 1 shows the key measurements made on the second set of patterns. The eight patterns (Fig. 3) were used as training samples, each shown to the vision system at least ten times in random positions and orientations in the field of view. A silicon vidicon camera, with a resolution of approximately 240 × 240 elements was used with a commercial vision system.[8] The field of view was 4in. × 4in. and the area of each pattern was π square inches. The system computed the mean values and standard deviations for each of the discriminating features used to differentiate and recognise each pattern. Shown in the table are the mean values and standard deviations for four selected features, namely, the net area, the hole area, the maximum radius from the centroid and the perimeter. Analysis of these data reveals the following conclusions:

[1] No **single** feature can differentiate all eight patterns. For example, the patterns in any subset consisting of two identical shapes except for one pattern having a hole, cannot be discriminated by the **maximum radius**, the **net area**, nor the

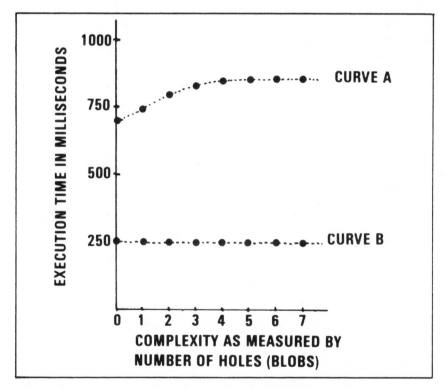

Fig. 4. Execution time as a function of complexity with constant shape and constant total area discs using the VS-100 vision system.

Samples: *2in. diam. discs with a variable number of holes from 0 to 7, each hole is ¼ in. dia. lying on a 1 in. dia. circle which is concentric with the disc.*

System and set-up: *Machine Intelligence Corp. VS-100 vision system with an MIC 22 camera having 128 × 128 elements. 50mm f:1.8 lens, 2.8 feet from object on light table. Field of view was 2in. × 2in. LSI-11/2 computer used.*

Discrimination features used for recognition: *Curve A: 11 features = total, hole, and net areas; perimeter: minimum, maximum, and average radii; major, minor axes of ellipse, some 2nd moments; length and width of bounding rectangle. Curve B: 1 feature = net area only.*

TABLE 1. DISCRIMINATION MEASUREMENTS								
Pattern		*Maximum Radius*		*Hole Area*		*Net Area*		*Perimeter*
No.	Shape	Mean	Standard Deviation	Mean	Standard Deviation	Mean	Standard Deviation	Mean	Standard Deviation
1	○	1.06	0.027	0	0	3.12	0.034	6.66	0.057
2	⊙	1.06	0.027	0.051	0.001	3.055	0.034	6.63	0.042
3	□	1.60	0.037	0	0	3.125	0.028	7.40	0.127
4	▫	1.61	0.037	0.055	0.0007	3.085	0.031	7.465	0.094
5	△	2.47	0.063	0	0	3.16	0.038	8.55	0.117
6	◬	2.43	0.053	0.054	0.001	3.06	0.023	8.46	0.114
7	▭	2.01	0.036	0	0	3.09	0.030	7.84	0.186
8	▭	2.00	0.031	0.055	0.001	3.049	0.025	7.86	0.175

perimeter features. The only feature able to discriminate for the presence or absence of a hole is the **hole area feature**. At the same time the **hole area** feature, by itself, cannot discriminate between the four subsets.

[2] The **net area** feature is not useful as a discriminating feature for all the patterns, since we have designed the experiment to maintain constant area for each shape, and the only differences are due to presence or absence of a single hole. The resolution of the system is too low to compensate for the noise and optical distortions which yield standard deviations comparable to the size of the ¼in. dia. hole. The major value, therefore, of this column of data is to inform the application engineer that more resolution elements are needed if this particular feature is to be of value for discrimination.

[3] The **maximum radius** feature is revealed as a useful discriminator between each of the four subsets. Together with the **hole area** feature, all eight patterns can be discriminated. For example, for the triangles and rectangles, the difference of the means, adjusted for three times the standard deviation spread is:

$$\underbrace{[2.47 - (3)\,(.063)]}_{\text{Max. radius-triangle}} - \underbrace{[2.01 + (3)\,(.036)]}_{\text{Max. radius-rectangle}} =$$
$$+\,2.281 - 2.118 = 0.163$$

This **positive** difference is $\dfrac{.163}{2.118} = 7.6\%$ of the measured means and is reasonably

TABLE 2. DISCRIMINATION FACTOR		
Discrimination between	*Maximum radius*	*Perimeter*
○ □	32.8%	2.8%
○ △	107.5%	20.5%
○ ▭	71.8%	6.8%
△ □	35.6%	5.6%
△ ▭	8.1%	−2.5%
□ ▭	11.9%	−6.7%

significant. Shown in Table 2 are the calculations for all the subsets using **maximum radius** and **perimeter** as the measured features.

[4] The **perimeter feature** is also a useful feature for discriminating three of the four subsets, but is marginal for discriminating between the square and rectangles. A calculation of the adjusted means indicates considerable overlap:

$$\left[\begin{array}{c}\text{Perimeter-rectangle}\\ 7.84 - (3)\,(.186)\end{array}\right] - \left[\begin{array}{c}\text{Perimeter-square}\\ 7.40 + 3\,(.127)\end{array}\right]$$

$$= 7.282 - 7.781 = -.499$$

This negative difference indicates considerable overlap and, therefore, this feature measurement is not useful as a discriminator between the rectangle and square. It is, however, useful for the other discriminations.

The combined use of **maximum radius** and **perimeter** measurements do, in fact, aid in assuring the discrimination of all the subsets, and the **hole area** measurements then can discriminate the two members in each subset.

Conclusions

A method for comparing some important elements of performance of machine vision systems is proposed and described. Based on the use of several sets of standardised, reproducible two-dimensional geometric patterns, execution time as a function of complexity, and discrimination capability can be compared. Initial experimental results are shown for one system.

References

[1] G. G. Dodd and L. Rossol, 'Computer Vision and Sensor-Based Robots', Plenum Press, New York (1979).

[2] C. A. Rosen and D. Nitzan. 'Use of Sensors in Programmable Automation', *Computer*, Vol. 10, No. 12 (1977).

[3] G. Gleason and G. J. Agin, 'A Modular Vision System for Sensor-Controlled Manipulation and Inspection', *Proc. 9th Int. Symp. on Industrial Robots*, Washington, DC, (March 1979).

[4] M. R. Ward, L. Rossol, S. W. Holland and R. Dewar, 'Consight – A Practical Vision-Based Robot Guidance System', *Proc. 9th Int. Symp. on Industrial Robots*, Washington, DC (March 1979).

[5] J. P. Foith, H. Geisselmann, U. Lubbert and H. Ringshauser, 'A Modular System for Digital Imaging Sensors for Industrial Vision: *Proc. 3rd CISM IFToMM. Symp. on Theory and Practice of Robots and Manipulators*', Udine, Italy (September 1978).

[6] J. P. Foith, 'A TV-Sensor for Top-Lighting and Multiple Part Analysis, *2nd/FAC/IFIP Symp. on Information Control Problems in Manufacturing Technology*, Stuttgart, Germany (October 1979).

[7] G. J. Agin, 'Computer Vision Systems for Industrial Inspection and Assembly', *Computer*, Vol. 13, No. 5 (May 1980).

[8] Model VS-100 Machine Vision System, Machine Intelligence Corporation, Palo Alto, California.

Chapter 3
RESEARCH

More effort is going into vision research than into any other subject of robotics but many problems still remain. Visual servoing, three dimensional viewing and illumination are some of the problem areas and the four papers cover these subjects.

ADAPTIVE VISUAL SERVO CONTROL OF ROBOTS

Arthur C. Sanderson and Lee E. Weiss, Carnegie-Mellon University, USA

Abstract

Visual servo robot control systems provide feedback on the relative end-effector position of a robot. They offer an interactive positioning mechanism which depends upon extraction and interpretation of visual information from the environment. In this paper, we characterise visual servo systems by the feedback representation mode, position-based or image-based, and the joint control mode, closed-loop or open-loop. The design problems posed by nonlinear and coupling transformations introduced by visual tracking systems are discussed, and a design strategy which utilises adaptive control, direction detection, and a logical control hierarchy is proposed.

Introduction

The interaction of manipulators with sensors has always been an important goal in the development of robotic systems. Such sensor-based systems would have increased functional capabilities as well as flexibility in the execution of tasks. Most practical systems which have exploited this sensor-robot interaction have focused on the development of communications links between existing sensing and robot systems. Such experiments have demonstrated the concept of interactive sensing but have not confronted the basic issues of analysis, design and performance evaluation of such systems. The principal objective of this paper is to describe a formal approach to the analysis of visual servo control structures and the potential use of such tools for the design and evaluation of visual servo systems.

A number of factors have delayed the practical development of sensor-based servo systems. Robot positioning systems and robot control systems are difficult to analyse and design in themselves, and there are a large number of practical applications of non-sensing robots in highly structured industrial and other environments. Sensing systems, particularly vision, are often slow relative to manipulator dynamics, and practical applications or machine vision are currently also limited to highly constrained situations. The analytical complexity of both manipulator control and sensory data interpretation make general formulation of the sensor-based control problem challenging.

In this paper, we attempt to isolate, as much as possible, the properties of the control problem which arise due to sensory feedback from those issues related to manipulator kinematics and dynamics or specific types of image processing and interpretation. In particular, we focus on the choice of feedback representation space and distinguish, for the case of vision, between image-based visual feedback which utilises image features as the control parameters, and position-based visual feedback which interprets the image in terms of Euclidean space positions and object geometry and utilises position and orientation parameters for control. Secondly, we have identified two basic classes of control structures. The 'look-and-move' structures utilise inner closed-loop joint control. The 'visual-tracking' structures have no closed-loop joint control and rely only on sensory feedback to drive the manipulator. The most novel of these systems are the image-based servo (IBVS) systems, and we discuss some approaches to resolving the complexities of a nonlinear, coupled feature transformation which may be of direct interest for control of the manipulator dynamics itself.

Background

A variety of experiments have been carried out on systems involving interaction of computer-based vision with robotic manipulators.[1, 2, 3, 4, 5, 6, 7, 8, 9] While several configurations of such systems are possible, most previous visual servoing applications in robotics have focused on the control of a robot arm or manipulator into a desired pose relative to a workpiece. One such configuration is shown schematically in Fig. 1 where the camera is mounted on a robot arm and information is fed back to position the robot end effector. In other instances, one might control the relative position of objects, the relative position of an arm and a fixed camera, or the trajectory of a camera mounted on a mobile robot.[10] Visual servo control of robot end-effector position with a fixed camera can be enhanced by structured visual design of the end-effector itself.

Robot positioning systems

A typical robot positioning system is shown in Fig. 2. The desired end-effector position of the robot may be specified by a homogeneous transformation T_6 (for a six-link manipulator) with respect to the base or equivalently by a concatenation of link transforms:

$$T_6 = A_1 A_2 A_3 A_4 A_5 A_6, \tag{1}$$

where A_i is the homogeneous transform relating position and orientation of the ith

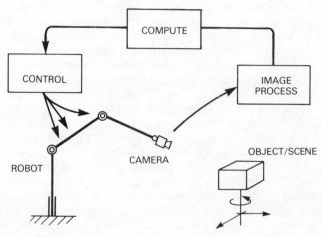

Fig. 1. Visual servo control.

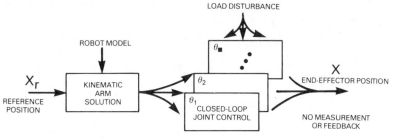

Fig. 2. Robot positioning system.

link to the i-1-th link. Transformation of a desired end-effector position to desired joint positions thus requires the solution of such a set of transcendental equations,[11] and may be computationally expensive. It is clear that the choice of the representational mode for the desired state of the manipulator may affect the dynamics of the system. In this case, the choice of end-effector position instead of joint positions will introduce time delays due to computation and noise due to inaccuracies of the arm solution.

While the homogeneous transformations in equation (1) are all linear transformations in the coordinate space, they may be highly nonlinear functions of the configuration variables and therefore yield complex equations. Such nonlinearities affect the open-loop system shown in Fig. 2 as a series gain terms and are normally considered separately from the design of the joint servo controllers. In this case, they do not affect the basic dynamics of the joint servo loops. In position-based visual servoing (or any other position-based end-effector feedback) the arm solution will reside inside the visual servo loop and may significantly affect closed-loop dynamics of the system. One useful description of the arm solution viewed as a configuration dependent gain matrix is provided by the Jacobian matrix, J:

$$d\vec{p} = J_{p_0} \, d\vec{q} \tag{2}$$

where $\vec{p}_0$ is some reference position and orientation of the end-effector, $\vec{q}$ is a vector specifying joint positions, $d\vec{p}$ is the differential positions and orientation, $d\vec{q}$ is the differential joint position, and J_{p_0} is a matrix of partial derivatives $\delta\vec{p}/\delta\vec{q}$. The Jacobian matrix may be viewed as the matrix of small-signal gains near a reference position $\vec{p}_0$. The terms of J are typically nonlinear functions of $\vec{q}$.

In the robot positioning systems shown in Fig. 2, the output of the arm solution is a set of joint trajectories. In practice, the joint controllers are designed using simplified dynamic models. Typically, position and velocity servo loops are designed toward critically damped responses on individual joints assuming decoupled links and some effective inertia and damping. The characteristic frequency of the servo loop relative to the structural resonant frequency of the manipulator and steady-state errors are also normally incorporated into the design. A variety of studies[12-16] have examined control strategies which take more detailed dynamic issues into account. Computed torque techniques[15] may be particularly useful in reducing system coupling.

Visual feedback

In visual servo control systems, information regarding relative object-camera position is available through processing of the image received by the camera. The image received is a two-dimensional array of light intensities which is not related in a trivial way to object position or geometry. Three types of computer processing are typically carried out on the image array:

1. **Image preprocessing** maps I to I_p, another two-dimensional array:

$$I_p = \quad (I), \tag{3}$$

2. **Image feature extraction** maps I_p to a one-dimensional array $\bar{f}$ or a relational feature description F:

$$\bar{f} = \quad (I_p). \tag{4}$$

3. **Image interpretation** maps $\bar{f}$ or F to a geometric model M of the perceived object and the position and orientation $\bar{r}_c$ of the object relative to the camera. In image-based feedback control we use $\bar{f}$ or F as the feedback parameters. In position-based feedback control we use $\bar{r}_c$ as the feedback parameters.

From the point of view of visual servo control, these image processing steps may be considered as observation or measurement processes on the end-effector position as suggested in Fig. 3. Such a measurement process affects the closed-loop system in several ways:

○ The observations add *noise* to the closed-loop system.

○ The observation procedures require time to compute and therefore introduce *time delays.*

○ *Reference parameters* must be supplied which correspond to the feedback parameters being used. In position-based systems, reference positions must be supplied, while in image-based systems, reference features must be supplied.

For the purposes of analysis here we will assume ideal feature extraction and image interpretation processes such that

$$E(\hat{\bar{r}}_c) = \bar{r}_c, \tag{5}$$

where E (·) denotes expectation value and $\hat{\bar{r}}_c$ is the estimate derived from image interpretation, and

$$E(\hat{\bar{f}}) = \bar{f}_G, \tag{6}$$

where $\bar{f}_G$ is the image feature vector derived from purely geometrical considerations of the object and its projection. We assume, in addition, that $\bar{f}_G$ is a smooth and

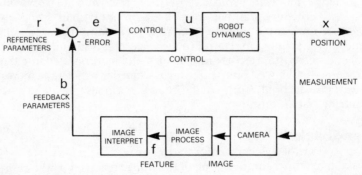

Fig. 3. Visual servo control system.

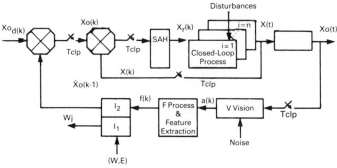

Fig. 4a. Static Look-and-Move.

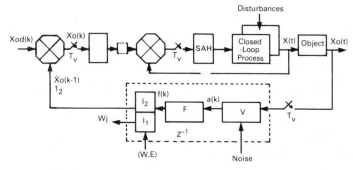

Fig. 4b. Dynamic Look-and-Move.

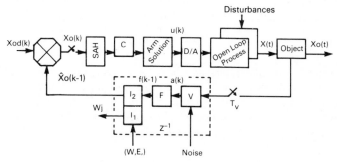

Fig. 4c. Position-Based Visual Servoing.

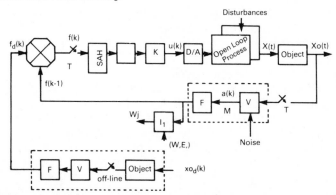

Fig. 4d. Image-Based Visual Servoing.

bounded (though not necessarily monotonic) function of relative object position and camera parameters, $\bar{s}_c$:

$$\vec{f}_G = \mathbf{R} \ (\vec{r}_c, \vec{s}_c). \tag{7}$$

The function $\mathbf{R}$ $(\cdot)$ may be derived analytically from a model of the object and the appropriate perspective transformation for the camera. Typical features including image areas, line lengths, and corner angles satisfy these relationships and are often non-monotonic.

In position-based visual servo control, the desired object position will be expressed in camera coordinates when the camera is moved or perhaps in world coordinates when the object is moved. In either case, transform relations such as

$$O = Z^z T_6 \ ^{T_6}C \ ^cO \tag{8}$$

may be used to relate position estimates in one coordinate system, to those in another. The feedback relations are linear and uncoupled (though perhaps statistically dependent) with added noise and time delay.

In image-based visual servo control, the feedback signal is non-linear and features may be coupled to several end-effector position parameters through the function $\mathbf{R}$. The image-based approach may reduce computational delays and estimation noise since interpretation is not required, and while avoiding the use of an arm solution in the forward path, the feature transformation itself may introduce non-linearities and coupling in the feedback path. The effect of the feature transformation may be viewed in much the same was as for the arm solution by calculating the Jacobian of the transformation. For clarity, in this case, we have called the Jacobian of the transform the *feature sensitivity matrix*, G, such that

$$d\vec{f} = G_{p_o} \ D\vec{p}, \tag{9}$$

where G is a matrix of partial derivatives, $\delta\vec{f}/\delta\vec{p}$ taken around the reference point p_o. The feature sensitivity matrix is therefore a small signal gain matrix in the feedback loop which is dependent on end-effector position and orientation, object position and orientation, and camera parameters. Since $\mathbf{R}$ $(\cdot)$ is not a monotonic function, we have also used the *feature direction matrix*, D,

$$D = \text{sig}(G), \tag{10}$$

to determine the direction of feature change. The off-diagonal terms of G determine the coupling between features and positions.

Visual servo control structures

Table I summarises the visual servo control structures which we have defined using various manipulator, feedback, and timing mechanisms. Corresponding block diagrams for the structures are shown in Fig. 4. The distinction between static and dynamic look-and-move structures is basically dependent on timing. In the static case manipulator control and image processing are completely non-coincident, while in the dynamic case both proceed simultaneously though perhaps at different sampling rates. The static look-and-move structure is not a genuine closed-loop control system since the feedforward and feedback dynamics are completely separated, but most experiments in visual servoing are carried out in this mode.

Image-based look-and-move (IBLM) systems require a feature-to-joint arm solution and will not be discussed in detail here.

TABLE I
SUMMARY OF VISUAL SERVO CONTROL STRUCTURES
MANIPULATOR

VISUAL	Closed-Loop Joint Control (Look-and-Move)		Open-Loop Joint Control (Visual-Tracking)
FEEDBACK Position-Based	Static Fig. 4a	Dynamic Fig. 4b	PBVS Fig. 4c
Image-Based	IBLM (Not shown)		IBVS Fig. 4d

Adaptive visual servo control systems

In order to focus attention on the issues introduced by sensor-based control, we have concentrated on visual-tracking servo control systems. Visual feedback introduces non-linearities, configuration dependence, and coupling in both the position-based and image-based systems. While one can design compensation for optimal system performance around a given reference point with average manipulator parameters, such designs will be optimal only in local regimes of the configuration space and compensation parameters should be changed for large motions.

Fig. 5 shows the results of a simulation experiment in image-based visual servo control of a three degree-of-freedom X, Y, θ manipulator with respect to a known object. The object is a pyramid and the perceived image is shown in Fig. 5a. Three features are measured in this example:

$f_1 = C_1 =$ centre of gravity of face 1,
$f_2 = A_1 =$ absolute area of face 1,
$f_3 = A_r =$ relative area of faces 1 and 4.

The feature sensitivity matrix calculated around this reference point $\theta = 50°$ is:

$$G = \begin{bmatrix} .084 & .000051 & .00015 \\ -.00014 & -.00024 & -.0023 \\ -.056 & -.0013 & -.92 \end{bmatrix}$$

It is clear from G that f_1, centre of gravity, depends most strongly on position x, f_2, absolute area, varies with both depth, y, and angle, θ, and f_3, the relative area depends primarily upon angle θ. An example of the step response of depth y to a unit step in desired feature f_2 is shown by the solid line in Fig. 5b. The plant in this case is a linear motor model and the visual servo loop proportional gain has been set to provide critical damping at this reference position, 50°. If the reference position of the object is changed to 80°, the visual servo system with the same proportional gain gives the response indicated by the dashed line in Fig. 5b. Clearly, the feature sensitivity matrix has changed and the previous design point no longer provides adequate response.

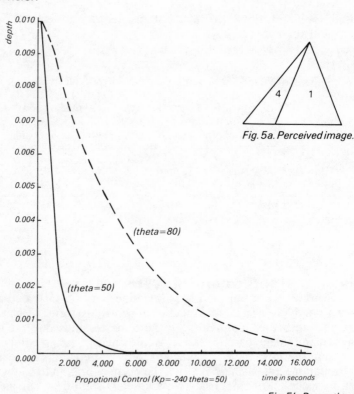

Fig. 5a. Perceived image.

Propotional Control (Kp=-240 theta=50)   *time in seconds*

Fig. 5b. Proportional Control
(Kp=-240 theta=50).

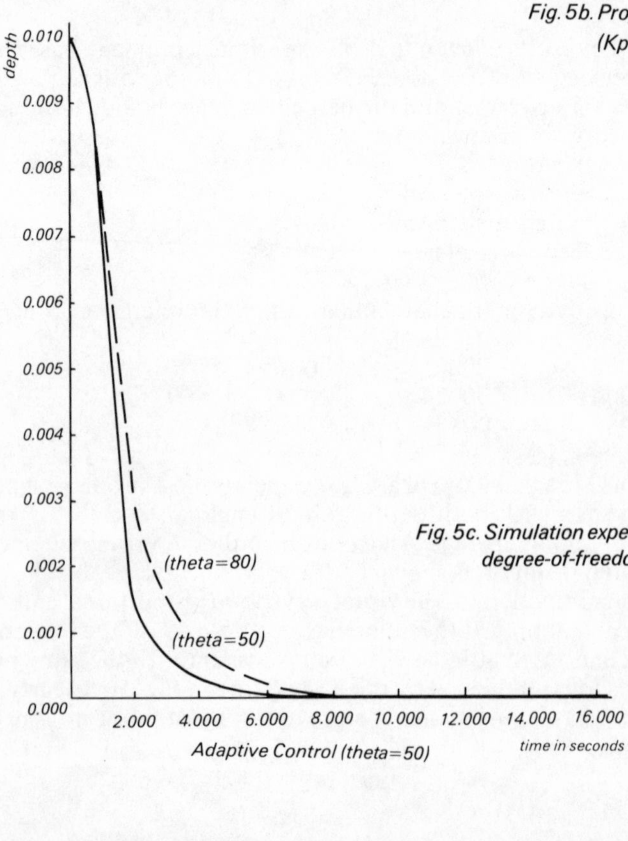

Fig. 5c. Simulation experiments with three
degree-of-freedom manipulator.

Adaptive Control (theta=50)   *time in seconds*

One approach to the problem of unknown or varying system parameters is to use adaptive control techniques which change the compensation parameters as system performance changes. Both model reference and self-tuning regulator adaptive control strategies have been investigated.[17, 18, 19] While formal results for stability and performance design are available only for linear systems with unknown parameters, adaptive controllers are used in non-linear and time-varying systems as well. The principal advantage of adaptive control in the present context is to enlarge the regime within which a given control strategy will provide good performance. A variety of adaptive control approaches are being examined for use in the visual servo problem. One self-tuning regulator design described in[9] was applied to the example above and the results are shown in Fig. 5c. The proportional gain of the system was adapted by the self-tuning regulator to maintain the dynamic performance close to its original specifications even though the system gain matrix changed with reference position.

While the adaptive controller compensates for gain variations due to feature coupling or joint coupling, it works well only in regions of monotonic feature dependence. The problems of direction, feature selection, and feature decoupling all require some supervisory control mode to adapt the structure of the controller as well as the parameters of the controller. In our current experiments, we utilise the D matrix as a map of monotonic feature dependence in order to determine the direction for control. In addition, we employ a *logical control hierarchy* as a mechanism for selecting features to be used in closed-loop control at a given time. The switching matrix S has terms b_i with values of 0 or 1 such that

$$\vec{f}' = S \vec{f}, \tag{11}$$

is a reduced feature matrix. The reduced feature sensitivity matrix

$$G' = S G, \tag{12}$$

then is adjusted to select features with desirable gain terms and reduce feature coupling. b_i is a logical variable whose value is determined by a set of syntactic rules:

$$\alpha_i \rightarrow b_i, b_j, b_k \tag{13}$$

where α_i is a logical expression based on tests of ongoing performance. In our current studies, the set of rules α_i is prioritised to avoid ambiguity in feature assignment. Simulations of three and five degree-of-freedom manipulators have shown that such *logical control hierarchies* are complex for arbitrary motions. Stable performance for such systems in limited ranges of motion has been achieved, and continuing studies are attempting to generalise the design of the hierarchical rules.

Conclusions

Visual servo robot control systems provide feedback on the relative end-effector position of a robot. They offer an interactive positioning mechanism which depends upon extraction and interpretation of visual information from the environment. While static look-and-move systems have been implemented using communications protocols between existing systems, significant challenges arise in the design of integrated sensor-based manipulator systems of the dynamic look-and-move and visual-tracking varieties. In this paper we have discussed the design problems posed by non-linear and coupling transformations in visual-tracking systems by the use of different representational modes. A design strategy which utilises adaptive control, direction detection, and a logical control hierarchy is being studied for use with these systems. Successful development may provide useful tools for the control of the manipulator itself where coupling and non-linearities are inherent in the dynamics.

Acknowledgements

This work was supported in part by the National Science Foundation under Grant ECS-7923893, and by a grant from the Westinghouse Electric Corporation. The authors would like to thank Dr. Charles Neuman for his helpful discussions.

References

[1] D. Nitzan and C. Rosen, et. al. 'Machine Intelligence Research Applied to Industrial Automation'. Tech. report 9, SRI International (August 1979).

[2] G. Agin, 'Real Time Control of a Mobile Robot with a Camera'. Tech. note 179, SRI International (February 1979).

[3] G. Vanderbrug, et. al. 'A Vision System for Real Time Control of Robots', *Proc. 9th Int. Symp. on Industrial Robots*, Washington, DC. USA, pp. 213-231, Society of Manufacturing Engineers (March 1979).

[4] S. Kashioka, et. al. 'An Approach to the Integrated Intelligent Robot with Multiple Sensory Feedback: Visual Recognition Techniques', *Proc. 7th Int. Symp. on Industrial Robots*, Tokyo, Japan, pp. 531-538, Japan Industrial Robot Association (October 1977).

[5] K. Tani, et. al. 'High Precision Manipulator with Visual Sense', *Proc. 7th Int. Symp. on Industrial Robots*, Tokyo, Japan, pp. 561-568, Japan Industrial Robot Association (October 1977).

[6] M. R. Ward, et al. 'Consight: A Practical Vision-Based Robot Guidance System', *Proc. 9th Int. Symp. on Industrial Robots*, Washington, DC, USA, pp. 195-211, Society of Manufacturing Engineers (March 1979).

[7] J. Birk, et. al. 'General Methods to Enable Robots with Vision to Acquire, Orient, and Transport Workpieces'. Tech. report 5, University of Rhode Island (August 1979).

[8] R. Abraham, 'APAS: Adaptable Programmable Assembly System'. *Computer Vision and Sensor Based Robots*, G. Dodd and G. Rossol, eds., Plenum, pp. 117-140 (1979).

[9] A. C. Sanderson and L. E. Weiss, 'Image-Based Visual Servo Control Using Relational Graph Error Signals'. *Proc. of the Int. Conf. on Cybernetics and Society*, Cambridge, MA, IEEE SMC, pp. 1074-1077 (October 1980).

[10] H. P. Moravec, 'Obstacle Avoidance and Navigation in the Real World by a Seeing Eye Robot Rover'. Tech. report 3, Carnegie-Mellon Univ., The Robotics Institute (September 1980).

[11] R. P. Paul, *Robot Manipulators: Mathematics, Programming, and Control*, MIT Press (1981).

[12] J. S. Albus, 'A New Approach to Manipulator Controls: The Cerebellar Model Articulation Controller (CMAC)', *Trans. ASME Journal of Dynamic Systems, Measurement and Control*, pp. 220-227 (September 1975).

[13] A. K. Bejczy, 'Robot Arm Dynamics and Control'. NASA – JPL Technical Memorandum, 33-669 (February 1974).

[14] B. K. P. Horn and M. H. Raibert, 'Manipulator control using the configuration space method', *The Industrial Robot*, Vol. 5, No. 2, pp. 69-73 (June 1978).

[15] J. Luh, M. Walker and R. Paul, 'Resolved Acceleration Control of Mechanical Manipulation'. *IEEE Trans. on Automatic Control*, AC-25, 3, pp. 468-474 (June 1980).

[16] D. E. Orin, R. B. McGhee, M. Vukobratovic, et. al. 'Kinematic and Kinetic Analysis of Open-Chain Linkages Utilising Newton-Euler Methods'. *Math. Biosc.* 43, pp. 107-130 (1979).

[17] Y. D. Landau, *Adaptive Control – The Model Reference Approach*. Marcel Decker (1979).

[18] K. S. Norendra, ed., *Applications of Adaptive Control*, Academic Press (1980).

[19] K. S. Norendra, ed., *Proc. Workshop on Applications of Adaptive Systems Theory*, Yale University (May 1981).

USE OF A TV CAMERA SYSTEM IN CLOSED-LOOP POSITION CONTROL OF MECHANISMS

P. Y. Coulon and M. Nougaret, Laboratoire d'Automatique de Grenoble, ENSIEG, France

Abstract
For industrial robots and automatic production, computer vision has been used for inspection and shape recognition. Applications for trajectography have also been reported. Not much work has been done on using a TV system as a position sensor to control, in the usual way of feedback control, the movement of a mechanism in a closed-loop position system. This paper gives an approach to this problem. Using an experimental apparatus, the static measurements have been studied, it is concluded that automatic calibrating is required. The dynamic characteristics of the measurement are experimentally studied and a theoretical model is given. Use of a TV-measuring system in closed-loop position control is then analysed and described.

Introduction
Using cameras for inspection tasks and robotics is rapidly increasing. While a great deal of work is available on the analysis of images of still objects, it must be noticed that the results on the problem of moving objects are still scarce[1, 2]: For practical applications this problem appears in three cases. A first case is trajectography digitisation (for instance following the path of an amobea under a microscope and analysing its movement). A second case is localisation and inspection of objects moving on a conveyor belt. A third one which is a synthesis of the two, is to use the camera as a position sensor in dynamic closed-loop position control of the movement of a servo-mechanism to reach a given region of space.

As the measurements must be made while the servo-arm is moving towards its aim, the effect of the speed of movement must be considered with two terms in mind, namely: what is the accuracy (velocity tracking error) and how is the stability of the closed-loop position control system guaranteed.

At the Laboratoire d'Automatique de Grenoble we have designed and built a computer vision system[3] oriented towards cinematographic and inspection measuring tasks. Equipped with two black/white, and one colour, TV cameras this system uses an LSI-11 (DEC) minicomputer. It is linked to a Norsk-Data computer and interfaced to microprocessors for special hardware (servo mechanisms with DC motors and stepping motors).

Using micro-x-y-rotation coordinate tables controlled by stepping motors, we have conducted many experiments to test the accuracy of the TV system for inspection and measuring tasks[4] for both still and moving objects. For moving objects, as stepping motors were too slow we have designed a low speed controlled belt, described below, and developed software for localisation and inspection of objects[5, 6]; in this paper we shall restrict ourselves to the problem of using the TV camera as a position error sensor in closed-loop control of mechanisms.

As usual, for sensors used in feedback loops, the important questions are of two kinds: static accuracy (linearity, fidelity, range) and dynamic aspect play a crucial part in the stability of a closed loop system.

System description

A system has been specially designed to locate a moving object and to yield its position both in digital and analogue forms. The analogue signal is very convenient to use for recording and in servo position control. Precise micro-x-y and x-y-θ coordinate tables have been used for calibrating studies. To analyse what happens when an object is moving a special apparatus has been designed to move a well contrasted object, at constant speed, in front of the camera. A brief description of the complete system is given here.

The position-measuring system[7]

It is built around a TV-camera (Grundig with a Resistron IND 2225 tube). The principle is quite simple, the object being supposed as well-contrasted with respect to background, a threshold is used to detect the object limits.

At the beginning of each TV line, a counter, CX, is reset to zero. This counter receives a 4 MHz clock signal which is counted as long as the threshold signal is not activated. When the threshold signal is present the CX counter is stopped. We get thus, in counter CX, a number which gives a measure of the position of the left side of an object with respect to an arbitrary zero column taken as the Y axis.

As soon as counter CX is stopped, a second counter CD is started, it receives then the 4 MHz clock and runs until the activation of the threshold signal which occurs

INTERFACE

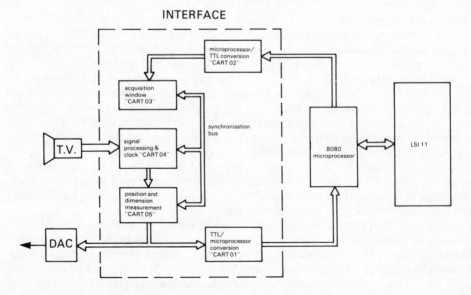

Fig. 1. Measuring system block diagram.

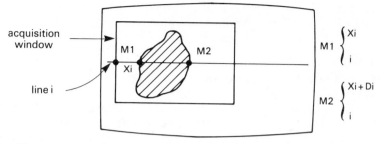

Fig. 2. Type of image segmentation.

when the right end side of the object is encountered. Every TV line period (64 micro-seconds) we get thus the position of the left side of the object (counter CX) and the width of the object (counter CD), for the current TV line.

A third counter, synchronised with the frame clock signal is reset to zero at the beginning of the frame associated with the upper left corner of the image. This counter CY, counts the lines, its information is used to indicate the line number where the object appears (this gives the Y position of the upper side of the object) and then disappears (Y position of the lower side of the object).

As a complete TV image is made of two interleaved frames with respectively 312 and 313 lines, every 20 milliseconds we get a 312×2 matrix. Each line of this matrix is given by the corresponding TV line. The first column is made of the counts delivered by CX. These counts are zero for all the lines where no part of the object was found, and they correspond to the X positions of the left end side of the object otherwise. The second column contains the successive widths of the object, as delivered by counter CD, along the successive TV lines.

By appropriate timing logic, a window can be chosen on the whole TV image. It permits to delimit the area of interest where the object is going to be searched using any a priori knowledge of its expected position. This window can be programmed to automatically track an object.

An 8080 microprocessor stores the 312×2 matrix delivered by the counter's systems. It serves as a buffer for an LSI-11 minicomputer. Moreover the 8080 is used to compute the coordinates of the centre of gravity of the object and its area.

Two D to A converters yield analogue measurement of these coordinates for recording or position servoing purposes.

With the 4 MHz clock for the CX and CD counters, the values of X and D are obtained as 8 bits binary words because each 64 microsecond TV line can be divided in 256 time increments given by the clock period of 0.25 microseconds.

The value of Y is more finely quantised because each frame is composed of 312 lines and thus Y is coded as a 9 bit binary word.

Fig. 1 presents the hardware block-diagram. Fig. 2 illustrates the analysis of an image, for the ith TV line, starting from the left edge of the window, counter CX measures X_i while counter CD measures the width D_i.

Mechanical apparatus

To study the influence of the speed of a moving object on the measurements, the apparatus shown in Fig. 3 has been developed.

A white belt with a black strip (the object) is run at a controlled speed by a DC motor.

The camera was placed two metres from the belt, the width of the belt is 50mm and its length is 1.5 metres. The black 'object' was a rectangle of various dimensions (from 10mm to 100mm).

The speed can be varied, in both directions between 0 and 5m/sec. The same system can be used to study closed loop postion control of the black piece on the belt through the TV position-measuring system.

Steady state considerations

The mapping: Object Plane – Electronic Plane

In the simplest situation the object to be located lies in a plane, let OP be this object plane. The optical vision system (mirror, diaphragm, lens) gives, in the GAUSS approximation, an image plane (IP) where the image of the object is found. The photo-sensitive electronic device intercepts the light cone produced by the optical system, and defines its own plane, let us call it the EDP (Electronic Device Plane). This EDP is the one which is read either by a fixed grid matrix (CCD matrix) or by an electronic scanning beam (in a vidicon TV system).

Conceptually these three planes must be considered to understand the various effects involved.

If the OP is orthogonal to the optical axis, then the IP is also orthogonal to this axis, in a good-manufactured optical system, the EDP also is perpendicular to this axis. The EDP and the IP can then be superimposed by focusing the optical lenses. The situation that we have just described is in reality never exactly realised and induces various errors.

Basically, we may consider that there is a mapping (in the geometrical mathematical sense) between the object plane and the electronic device plane (which yields the processed and computed signal). Ideally speaking this mapping must be known to deduce the actual movement of an object from its computed position as obtained from the EDP signal. Two situations are encountered: trajectography (or position measurement) and error position measurement (for servoing purpose), and both lead to different requirements.

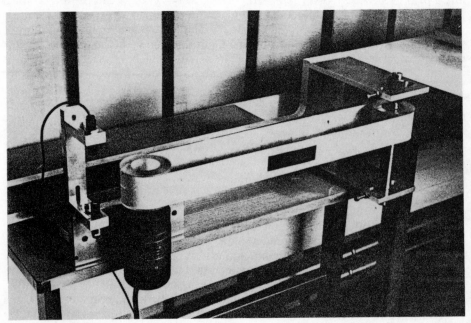

Fig. 3. Controlled speed belt conveyor.

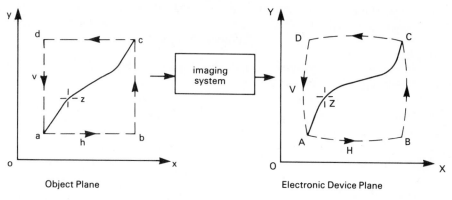

Object Plane   Electronic Device Plane

Fig. 4. Mapping OP—EDP.

Trajectography

In this problem, the position of a moving object must be measured with respect to a coordinate system situated in the OP. If the range of the movement is restricted enough a fixed lens and camera can be used (if it is not the case, the focal distance and orientation of the camera can be changed).

This case is illustrated by Fig. 4. The object is moving along line '*l*' while in the EDP, the line 'L' is described.

Let $Z(X,Y)$ be the position in the EDP, while the object is located at $z(x,y)$. The whole (geometrical, optical, electronic, coding) system produces the mapping

$$Z = f [z]$$

Ideally speaking, this mapping gives a simple relationship z and Z, such as: a grid in the OP yielding a grid (of reduced dimension) in the EDP.

Unfortunately this is not the case, the usual mapping having the barrel form sketched of Fig. 4.

The various phenomena involved in this mapping are (apart from image quality):

○ *The parallax effect:* when the object plane is not orthogonal to the optical axis; particularly if a mirror is used, or when the object is out of axis.

○ *Geometrical aberrations:* increases when the object is located away from the optical axis.

○ *The out of focus effect:* a point object yields an image with an area.

○ *The depth of field effect:* with 3D objects.

○ *Electronic drift and coding:* spatial sampling due to the coding grid, various drifts in the coding systems.

While all these problems can be reduced with good design practice it must be noted that they cannot be eliminated completely. Thus a simple and effective way of taking care of the problem is to introduce an automatic self calibrating system.

The mapping $Z = f(z)$ can thus be memorised. For instance, the object (using an x-y micro mechanism) can describe lines 'v' and 'h' (see Fig. 4), the associated curves 'V' and 'H' are tabulated. Any position Z being thus measured with respect to the curved coordinates system (which describes the mapping).

A simple way is to have reference points like a, b, c, d; to compute and store their measurements in the EDP points [A, B, C, D].

To compensate for drift and small changes in position (inclination of object plane, small movement of camera) the self calibrating procedure must be repeated either regularly or each time it is suspected that a significant change has occured in the environment. Moreover, the reference points may also be used for auto-focusing of the optical system.

We have used this kind of approach for measuring teeth profile on screws[6] and for trajectography of small (0.1mm) organisms ('ROTIFERES') living in water.[8]

Position Error measurement
Typically this case arises for instance when a robot arm with a gripper must move to take an object; as shown on Fig. 5. If we assume that the object and the gripper are in the object plane, the imaging system must give information not on the absolute positions of the gripper and the object but on the position error between the gripper and the object.

Moreover, this position error information does not need to be very accurate as long as it is sufficient to provide convergence of the zero-error seeking control algorithm which actuates the movement of the robot arm.

It must be emphasised that while the precise mapping between the object plane and the electronic device plane was required for trajectography, in servoing a robot gripper on an object, this requirement is far less stringent.

While this type of error-detector and zero-error seeking system are well known in servo mechanisms theory we may illustrate their advantage by two features.

Let: $Z_o = f(z_o)$ – the mapping of the object position, and
$Z_a = f(z_a)$ – the mapping of the arm position; in position-error sensor, the output information is D where
$$D = Z_o - Z_a = f(z_o) - f(z_a)$$

Thus, if any additive disturbance modifies the whole mapping (i.e. the imaging system) as both $f(z_o)$ and $f(z_a)$ are equally disturbed, the position difference D, remains the same (this is not the case for the absolute measurement).

Assume now that the mapping is affected by a multiplicative disturbance (change of gain in amplifier, out of focus, etc.) which is felt all over the field of vision.

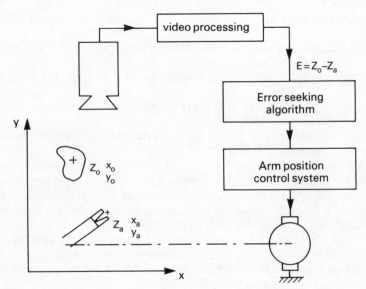

Fig. 5. Position error measurement.

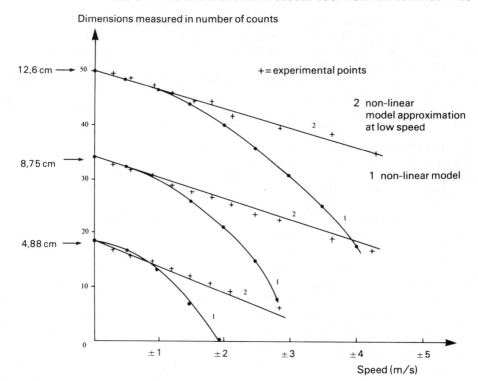

Fig. 6. Width variation with belt speed.

Thus
$$f(z) \rightarrow G(\tau)f(z)$$
The Position Error signal becomes:

$$D_{\text{modified}} = G(\tau)f(z_o) - G(\tau)f(z_a)$$
$$= G(\tau)[f(z_o) - f(z_a)] \tag{1}$$

Equation (1) shows that only the gain of the system is changed, this may eventually yield a change of the rate of convergence in a zero position error seeking algorithm but the modified position error signal remains equal to zero only when z_o = z_a (which insures that the algorithm still converges towards a zero error physically putting the gripper on the object).

These two basic features (no influence of additive or multiplicative disturbances) are the most interesting for servoing applications.

Dynamic analysis
Experimental results
Width measurements. For different sizes of rectangular black objects we have stored their widths, as given by the TV measuring system, with the conveyor belt running at constant speed. For these experiments the movement is parallel to a TV line. Fig. 6 shows the apparent sizes of objects as a function of speed. It is observed that the object seems to shrink when speed increases, almost linearly. This effect is symmetrical for the belt running in the same direction as the TV line electronic spot, or the reverse.

Positions measurements. It is observed for position measurements that the measured position always lags the true one; this velocity error increases with speed. If only the position of one side of the object is measured, the error is not symmetrical and it is not the same when the object moves in the same direction of the TV line spot analyser or in the opposite direction.

This unsymmetrical effect yields a distortion when the object is moved sinusoidally. To test the harmonic response of the TV measuring system, we have built a simple mechanical system to obtain a sinusoidal projected movement in front of the camera. Detailed experimental results are found in [7].

Mathematical model

To understand and explain the experimental results we have analysed all the different operations involved in the measuring process. First we have observed the analogue TV video signal. For a stationary object, the video signal is a fixed position pulse with steep transitions for a well contrasted image. For a moving object, the analogue signal is no longer a pulse but it is deformed and moves along the line. By taking pictures of these deformed pulses we have been able to develop a mathematical model which is in good agreement with the experimental results. The model uses the basic considerations that follows.

TV tube memory effect. The photo-sensitive material of the vidicon-tube can be considered as being made of independent resistive/capacitive cells. Each of these cells receives electric charges from the electronic beam of the TV spot and loses electric-charges when they receive light. Thus the intensity needed by a cell when the electronic beam charges it depends on the previous amount of light it has received. The intensity of the beam is amplified and thresholded in our measurement system.

The resistive/capacitive nature of the material results in a memory effect which can be described by a curve giving the fraction of the residual signal remaining after a given number of frames. This curve, given by the TV tube manufacturer, can be approximated by an exponential decay with a time-constant τ.

Video signal produced by a moving object. Let

E(x,t) : Distribution of light produced by a 1D object moving along the x axis.

U(x,t) : Electric charge, at time t, of the photo sensitive cell situated at x.

I(x,t) : Video signal, at time t, given by cell x.

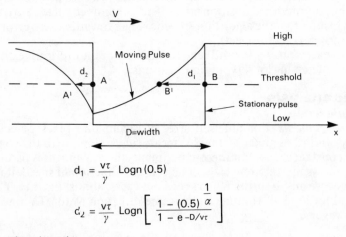

$$d_1 = \frac{v\tau}{\gamma} \, \text{Logn} \,(0.5)$$

$$d_2 = \frac{v\tau}{\gamma} \, \text{Logn} \left[\frac{1 - (0.5)^{\frac{1}{\alpha}}}{1 - e^{-D/v\tau}} \right]$$

Fig. 7. Deformed moving pulse.

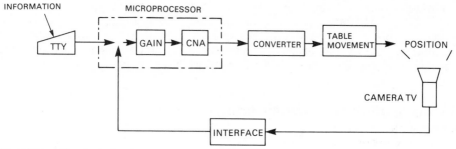

Fig. 8. TV position control system.

We have

$$I(x,t) = I_o + a \cdot U(x,t)^\gamma \quad \text{and} \quad U = bE$$

(I_o is called the 'dark current'; γ is constant for a tube; a and b are gains).

Due to the tube memory effect (time constant τ), we get:

$$U(x,t) + \tau \frac{d}{dt} U(x,t) = b E(x,t) \tag{2}$$

For a moving object, in a constant illumination, the distribution $E(x,t)$ moves along with speed v. Thus:

$$E(x,t) = f(x - vt)$$

Where $f(x)$ is light distribution of the stationary object, with Laplace transform $F(p)$.

Using standard Laplace transform technique equation (2) can be solved with intitial distribution $U(x,o) = O$.

We obtain then, for $U(x,t)$, a distribution which moves along the x axis with speed v.

Or

$$U(x,t) = g(t - \frac{x}{v})$$

with

$$g(t) = \mathcal{L}^{-1} [\frac{1}{v} \frac{F(-p/v)}{1 + \tau p}] \tag{3}$$

Application to rectangular objects. For a rectangular object, the light distribution received on the photo-sensitive material is a pulse which has a width proportional to the width of the object (determined by the ratio of object to image distances of the optical system). This pulse moves along with speed v. Function $g(t)$ given by equation (3) is the response of a time constant to a pulse input. The result is shown on Fig. 7.

If we consider now, a threshold, it is seen that for a stationary object the threshold is obtained in A and B, while for the deformed signal of the moving object the threshold is found in points A′ and B′. Assuming that the threshold have been set at the middle of the range for the stationary object, we compute easily d_1 and d_2 as shown on Fig. 7.

The apparent width reduction is obtained as $(d_1 - d_2)$.

The apparent centre of gravity lags behind the true one with an error

$$\frac{(d_1 + d_2)}{2}.$$

It must be noticed that in our earlier experiments we were measuring the position by the detection of only one side of the object; we observed a non symmetrical effect with positive or negative speed, this is because d_1 and d_2 being different, the TV spot encounters first A' or B', according to the direction of movement. A symmetrical error is obtained when the centre of gravity is computed.

TV controlled servo positioning

TV sensor transfer function

From the study developed above, we may give general results on the use of TV as a sensor in servo position systems. For very slow movements, the memory effect can be neglected; to know the values of speed where the dynamic effect is negligible equations (2) and (3) may be used. If this is the case, as a position error senses for aim and arm belonging to the same plane, a preliminary calibration can be made but in a zero seeking algorithm it is not crucial. Dynamically, for slow movements only the sampling period (20m sec) of the sensor needs to be considered.

In fast moving servo-mechanisms, care must be taken. It is necessary to put a distinctive pattern (circular or rectangular on the moving arm) because the video signal deformation depends both on the pattern and on the memory effect curve of the TV camera. If a pattern too small in dimension is used, the threshold will give no signal at all if the speed is fast (see Fig. 7).

The formulas for d_1 and d_2 that are given, have been experimentally found satisfactory but they are valid only if the TV tube is used in its linear range.

For position measurements, considering that the velocity error is proportional to the speed v, a simple model is to consider that a pure delay gives a good approximation, and it is obtained as $TR = (d_1 + d_2)/2v$.

A practical example

To illustrate the use of a TV sensor we have done an experiment in which a servo-controlled axis is under camera control for reaching a specific point along the axis.

The servo-controlled axis was the X axis of an X-Y analogue recorder. This servo-mechanism was a second order system with a damping factor of 0.6 and a natural angular frequency $\omega_n = 12.5$ rad/sec. Measuring the position of the Y arm (which translates along X) with the camera system, a second position control loop was established through the camera and a microprocessor controlling the X input command of the X-Y recorder. The camera was modelled by a delay T_R; a proportional and integral controller was designed using the Z transform transfer function. Fig. 8 gives a block diagram of the experiment and Fig. 9 shows a step response of the closed-loop TV position control system. This demonstrates the role of the dynamic analysis of the TV sensor and its use in dynamic closed-loop position control.

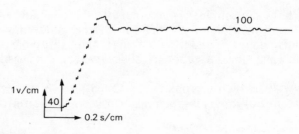

Fig. 9. Position step response.

Acknowledgement

The study has been conducted with a C.N.R.S. grant (project A.R.A.) whose support is kindly acknowledged.

References

[1] A. Gilbert, et. al. 'A real time video tracking system', *IEEE Trans*, Vol. PAMI, No. 1 (January 1980).
[2] G. Hirzinger, et al. 'Automated TV tracking of moving objects', *IEEE 5th Conf. on Pattern Recognition*, Miami (1980).
[3] J. P. Charras, 'Système intéractif d'acquisition d'images'. Thèse D.I., I.N.P.G. (July 1980).
[4] T. Redarce, 'Etude et réalisation d'une station automatique d'orientation', D.E.A., I.N.P.G. (July 1980).
[5] S. Olympieff, 'Reconnaissance de formes par caméra TV', D.E.A., I.N.P.G. (July 1980).
[6] P. Horaud, 'Automatic inspection and orientation of external screws' *IEEE 5th Conf. on Pattern Recognition*, Miami (1980).
[7] Y. Tahon, 'Localisation et mesures dynamiques par Vidicon', Thèse D.I., I.N.P.G. (May 1979).
[8] P. Y. Coulon, 'Trajectographie automatique de Rotifères', Rapport interne L.A.G. (July 1981).

TOWARDS A FLEXIBLE VISION SYSTEM*

*O. D. Faugeras, F. Germain, G. Kryze, J. D. Boissonnat, M. Herbert, J. Ponce,
E. Pauchon and N. Ayache,INRIA Domaine de Voluceau-Rocquencourt, France*

Abstract

A Vision System designed for building accurate models of industrial parts is
described. Potential applications include tolerancing testing, data base acquisition
and automatic recognition of objects. The system is made of a laser rangefinder that
measures the position in space of points on the parts by active stereoscopy, a table
on which the parts are positioned and can be translated vertically and rotated under
computer control, and a set of algorithms to produce accurate geometric models of
the part based on the measurements made by the laser. Representation and
recognition results are presented on a variety of objects as shaded graphics displays.

Introduction

With the advent of fast and reliable hardware for acquiring range data and the ever
more pressing need for sophisticated vision systems in automated assembly,
researchers have been faced with the problem of building accurate and efficient
models of 3-D objects to be used in recognition and manipulation tasks.

This paper discusses current research trends at INRIA in the field of 3-D range
data acquisition, modelling and object recognition. The acquisition system consists
of a computer controlled laser rangefinder and moving table on which objects are
positioned. Various approaches have been explored for the automatic production
of geometric models from the range data. These models are intended for data base
acquisition and automatic recognition of objects. Later we show how one of these
models can be used to recognise industrial parts from any viewpoint.

The acquisition system

The system which has been developed[1] is composed of a laser range finder, a
system of cameras, a computer controlled table and a microprocessor system.

General principle

The sensor used provides the z coordinate of a point on the surface of an object as a
function of the x corordinate (Fig. 1). The basic principle is that of active
stereoscopy. *Stereoscopy* because it uses at least two cameras to yield images from
different viewpoints and *active* because the difficult problem in stereo vision of

---

*Expanded version of a paper with the same title presented at the 12th international Symposium on Industrial Robots, 9-11 June 1982,
Paris, France.

matching the two views is avoided by over illuminating one point of the surface to be observed.

This strong lighting is provided by a laser beam which can be moved in the x direction. The deflection system is computer controlled and this makes the sensor random access. The laser beam creates a small spot on the surface which forms by diffusion a secondary source of light which yields one image on each camera. From the position of these images on the retinas, and the geometrical parameters of the cameras, it is possible to compute the x and y coordinates of the brightly lit point.

The platform on which the object rests is equipped with stepping motors. These motors raise and lower the table (this is the y-axis) and also rotate it (around the y-axis). Objects in size up to 750 × 750 × 600mm can be accommodated. The redundancy introduced by this rotation with respect to the sweeping of the laser beam allows us accurately to analyse concave objects.

The existing system

The principle of the sensor is very simple. The realisation is more tedious and requires the design and coordination of three distinct systems: the cameras, the laser and the logical unit.

The optics of the laser has two functions. First the beam must be scanned in the x direction. This can be done by means of an opto-electronic device, but a system composed of a mirror driven by a galvanometer is preferred. This is the only moving part of the whole apparatus. Second, the beam must be focussed. The optics are arranged so as to maintain the diameter of the laser beam constant and minimal in the whole spatial range of measurements. This arrangement is also a function of the laser physical parameters.

Each camera is associated with a microprocessor which determines the significantly lit diodes and performs an interpolation to obtain a position accurate to $\frac{1}{8}$th of a diode width. A second microprocessor controls the whole system, in particular the galvanometer, and performs the triangulation from the data provided by each camera.

This sensor is therefore relatively autonomous. The connection to a computer is implemented through the second microprocessor. Exchanges between the two are very limited. The computer sends the sensor the direction of the beam and receives the two coordinates of the measured point. In our current implementation, the measurement time is approximately 1ms.

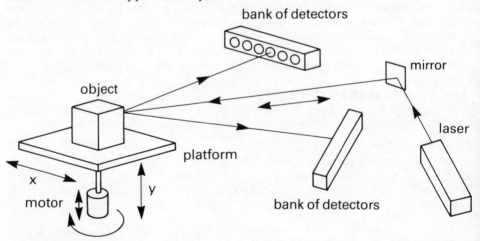

Fig. 1. The acquisition system.

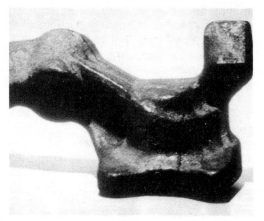

Fig. 2. A photograph of an automobile part.

Gathering 3-D data

This data is in the observer centred coordinate system. While creating a 3-D model of the object, an object centred representation is required. This is computed by marking the zero position for the x- and y-axis and obtaining a reference value for the z-axis of the platform on which the object rests (Fig. 1). Thus all the points can be transformed to the same coordinate system. While acquiring the data related to an unknown view of the object, the actual position or orientation of the object on the platform does not matter.

As an example, the complicated casting of an automobile piece is shown in Fig. 2. In order to create a 3-D model of the object, a range data image is produced for every 30° rotation of the object around the y-axis. Finally, top and bottom views of the object are taken. These two views were put in correspondence with the other views by using several control points on the object to compute the transformation. One of these views, obtained with the range data acquisition system described above is shown in Fig. 3. After thresholding the background points, each individual view had approximately 2,000 points. Surface points for the composite object were obtained by transforming the coordinates of the points in any of the views with respect to some common coordinate system.

Another way of gathering 3-D data with the existing system is to combine the sweeping of the laser beam with the rotation of the platform in order to digitise the contour of a plane section of the object; by lowering the platform one can digitise another such contour, and so forth. At the end of this procedure one gets a set of parallel plane section contours of the object.

At the present time the digitisation of one contour is entirely automated, provided that the contour has neither holes nor too deep cavities.[25] The system has been made interactive in such a way that the digitisation process stops when one of the above conditions is not fulfilled and the operator can easily show the system where it must resume the analysis of the contour. Once a set of such contours has been obtained, it is possible to construct an approximation of the surface between the contours, by using a graph theoretic approach to search for optimal tile arrangements, and to visualise this triangulation of the surface (Fig. 4) in order to control whether it is necessary or not to insert or to delete some contours in the set already obtained.

These data can be used to construct high order geometric models of the object, either by triangulating the points obtained, or by deriving a volumetric discretisation of the object from the contours (see below). But a first and direct application is, given the contours of successive horizontal cross-sections of an object, to identify and precisely locate this object standing in the same stable position on a horizontal plane. This is done by measuring the contours of a cross-section at a given height and matching these against those of the pre-stored horizontal cross-sections. This yields the horizontal rigid transformation and also the vertical translation undergone by the model, and eventually the spatial location of the observed object.

In fact, for most applications, it would be awkward and time consuming to rotate the observed object to measure the entire boundary of a cross-section. In this case a single sweeping of the laser beam provides only portions of the cross-section boundary, missing all the rear part of the object, and possibly some concave sections of the front part also. Therefore we have had to develop original algorithms dedicated to this partial matching problem,[26] and the results obtained are very promising. When not enough points are measured at a given height, the quality of the result may be improved by measuring some additional cross-section contours, at distinct heights, but this is done at the expense of the computing time. An illustration of experimental results obtained with real data, using the measurements taken from a single partial cross-section contour of the previously mentioned workpiece is shown in Fig. 5.

This technique is a good alternative to some more classical methods, relying upon the measure of the object silhouette by an above-located video system, when the environmental configuration prevents any vision from above, when the lighting conditions are too poor, or when the parallax become untractable.

Geometric models production

Our goal is to find suitable representations for recognition and description tasks. 'Representation' is defined as the act of making a description of an object. The description should capture an intuitive notion of shape which should be compact and simple and our attention was first directed towards the construction of surface models from range data. The work in this field is already important. Binford[3] proposed the concept of a generalised cylinder to represent 3-D objects. These are defined by a 3-D space curve, known as the axis, and a cross-section of arbitrary shape and size along the axis. Agin and Binford,[4] Nevatia[5] and Bajcsy and Soroka[6] all have used this concept to segment complex objects. Marr[7] proposes a hierarchy of models using generalised cones as a set of primitives. For a good review of existing models see [8]. Our approach is to use a structural description in terms of

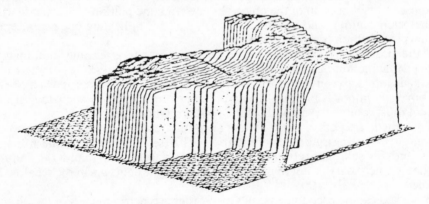

Fig. 3. Range data for the part of the Fig. 2.

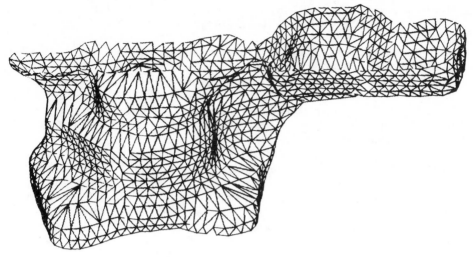

Fig. 4. Triangulation of the automobile part from the contours.

geometric primitives believing that these primitives must be easily extracted by a vision system. This is not the case of the generalised cones which is why we have used planar primitives and simple non-planar primitives such as cones and cylinders. A similar approach is used in [8-10]. More recently, we have investigated another description, viz. volumic models, which are easy to extract from our data.

Planar approximations

Three-point seed.[11] This first technique approximates objects by a set of planar faces and it is a two-stage process. First, a set of points that belong to various faces of the object is found using a three point seed algorithm and in the second stage, approximate the face points obtained by polygons. The three point seed method for the extraction of planar faces from range data is a sequential region growing algorithm.

In a well-sampled 3-D object, any three points lying within the sampling distance of each other (called a 3-point seed) form a plane (called the seed plane) which either

(a) coincides with that of the object face containing the points, or,

(b) cuts any object face containing any of the three points.

A seed plane satisfying (a) results in a plane from which a face should be extracted, while a seed plane satisfying (b) should be rejected. Two simple conditions that suffice to determine if a plane falls into category (b) are convexity and width.

The algorithm involves the following steps:

1. From the list of surface points select three points which are non-colinear and close, relative to the sampling distances.
2. Obtain the equation of the plane passing through the three points chosen in step 1.
3. Find the set of points P which are very close to this plane.
4. Apply the convexity condition to the set P to obtain a reduced convex set P'. This separates faces lying in the same plane.
5. Check the set P' obtained in step 4 for width.

6. If the face is obtained correctly (i.e. The convexity and width conditions are satisfied), remove the set of points belonging to this face from the list of points and proceed to step 1 with the reduced number of points in the list.

After the surface points belonging to a face have been obtained using the 3-point seed algorithm, two checks are made.

1. All the points which have been previously associated with various faces are checked for possible inclusion in the present face.
2. The set of points in the present face is checked for possible inclusion in previous faces.

The application of the above two tests provides the points which belong to more than one face. This information in turn provides knowledge about the neighbours of a face and relations among them which are in turn useful for object recognition tasks.

Now the surface points have been associated with various planar faces. Although some edge points and vertices will be known, an independent step is required to obtain polygonal faces. The polygonal approximation of a face involves the following steps.

1. Get the points belonging to a face.
2. Obtain the binary image of the face points.
3. Trace the boundary of the image obtained in step 2 using a boundary follower.
4. Perform a polygonal approximation of the boundary of the face.

The complexity of the 3-point seed algorithm is $O(n^4)$ where n is the number of data points.

The 3-point seed method was applied to the individual views shown in Fig. 3 and

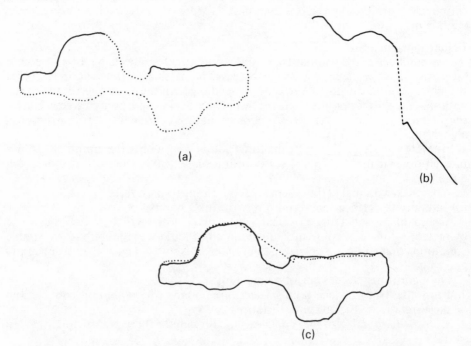

(a)

(b)

(c)

Fig. 5. Recognition of a contour of the part of the Fig. 2.
 (a) Total cross-section boundary at height 81 mm (109 segments).
 (b) Partial cross-section boundary (same height, $\theta = 52°$, 32 segments).
 (c) Best estimation of the matching transformation ($\theta = 51.7°$). Matched segments are solid lines in (a) and (b).

Fig. 6. Planar faces obtained with the 3-point seed method on the view of Fig. 3.

to the composite object. Fig. 6 shows the faces found for the 0° view (Fig. 3) of the object. In this figure various faces are shown in different shades of grey. As can be seen, most of the faces found are reasonable. The object has major curved surfaces that are split into different faces. Below it is shown how this type of representation can be used for object recognition.

Triangulation of objects. This section is devoted to the problem of building a polyhedral representation of an object digitised by the laser range finder. More precisely the problem is to find a regular polyhedron (i.e. one whose faces are triangles) whose vertices are the measured points. Notice that, because different views are necessary to cover the whole object, the measured points can not be ordered easily and are supposed, in the sequel, to be irregularly scattered throughout a domain in 3-D space.

The corresponding 2-D problem is the triangulation of a set of n points in the plane. This problem has received a lot of attention and the key result is that 0(n log n) is a lower bound to triangulate n points. For, a lower bound of the complexity of any algorithm which constructs a triangulation of a surface is 0(n log n). One of the most well-known triangulations is the Delaunay triangulation[12] which can be performed in 0(n log n) where n is the number of points.

In the 3-D space, the Delaunay triangulation can be directly applied when the surface can be projected injectively on a plane. Moreover, optimal triangulation is also possible when the surface is known to be convex. In that case, an optimal triangulation obtained for the unique convex triangulation is the convex hull of the set of points which can be computed in 0(n log n) time.[13]

The basic idea of our approach is to reduce the general case to these two cases. In order to achieve this use is made of the spherical representation of surfaces. This representation associates to a point M of the surface, the end-point of the normal of the surface at M. So the image of a surface lies on the unit sphere (the Gaussian sphere), which is, of course, convex.

The important fact is that the spherical representation of sufficiently small regions that are everywhere cup-shaped (called surfaces of positive curvature), or everywhere saddle-shaped (called surfaces of negative curvature), is one-to-one;[14] and so a triangulation on the unit sphere corresponds to a triangulation on the surface. In the case of surfaces of zero curvature, the projection on the tangential plane provides the triangulation.

The algorithm 'grows' the triangulation by propagating the contour of the triangulated domain. This propagation is done by looking around an edge of the contour which point of its neighbourhood must be taken into account to create a new triangle either on the spherical image of the surface or on the tangential plane according to the current sign of the curvature. The process stops when the frontier between two domains with distinct signs of curvature is reached. At this level, we can cross this frontier and continue the process on the next domain.

The process is initialised by taking the smallest edge of the convex hull which is the first edge of the triangulation.

Since this algorithm essentially chooses the best point, among a few points, to be associated with an edge, the complexity is the complexity of the search of the neighbours of an edge. Many fast algorithms have been proposed which perform such a search in 0(nlogn) time (see for example [15]). The complexity of our method is also 0(nlogn). The result of the triangulation of a simple object is shown in Fig. 7.

Polyhedral approximation of objects.[16] The technique described in the preceding section provides us with a triangulation of the object. This triangulation can be used for various purposes but in particular for obtaining a cruder polyhedral approximation. This approximation can in turn be used more efficiently as a model of the object.

The triangulation is a graph G = (V,E) where the set of vertices V is the set of points on the object and E is the set of edges. In our case the set of edges coincides with the set of triangle edges. We use this graph representation recursively to split the set of points and approximate each subpart independently by a polyhedron with triangular faces.

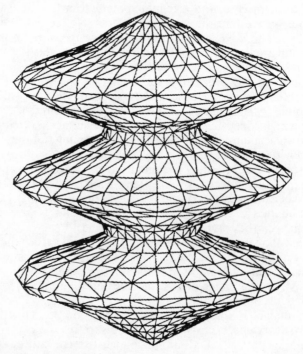

Fig. 7. Triangulation of a lamp-shaped object.

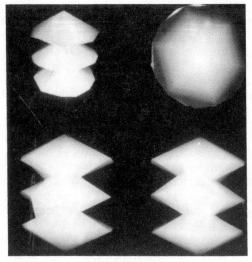

Fig. 8. Polyhedral approximation of the object of the Fig. 7.

The algorithm is as follows:
1. Pick three points P, Q, and R on the surface S which are not neighbours. The plane PQR is the zero order approximation of S.
2. Find the shortest cycle passing through the points PQR which is closest to their plane. This forms a cycle in the graph G which we denote by (PQR). Because G is planar, this cycle separates the surface into two surfaces S_1 and S_2 and the graph G into two disconnected subgraphs G_1 and G_2.
3. Process G_1 and G_2 independently: find the most distant point S_i to the plane PQR (i = 1,2). The polyhedron $(PP_1Q,QP_1R,RP_1P,PP_2Q,QP_2R,RP_2P)$ is the first order approximation of the surface S.
4. Then, recursively split the surfaces S_1 and S_2 into three as follows. Find the cycle (PP_iQ) in G_i which does not contain R. This cycle is chosen in such a way that the path PP_i, for example, is the shortest in terms of a set of weights equal to the distances of each point to the bi-sector of the planes PP_iR and PP_iQ. This splits G_i into G_{i1} and G_{i2}. Suppose G_{i2} is the component of G_i that contains R. We find the cycle (QP_iR) in G_{i2}. This splits in three ways: the graphs G_{i1}, G_{i2} and G_{i3} corresponding to the triangles PP_iQ, QP_iR and RP_iP, respectively, (i = 1,2).

Each subgraph (subsurface) is then processed independently in the same way as in 3. until an acceptable error level is reached. The net result is an approximation of the original surface S by a polyhedron whose faces are triangles. An undesirable property of this simple algorithm is that once an edge of the polyhedron has been created, it can never disappear even though it may be quite a bad approximation to the surface. We therefore refined the algorithm in a simple way to allow for breaking of edges and thus approximate the surface better. The complexity of this algorithm can be shown to be $0(Tr.n.\log(n))$ where Tr is the final number of triangles and n the final number of points. Since it requires an initial triangulation as its input, we must add the complexity of the triangulation of objects method described above which is also $0(n.\log(n))$. This is thus more efficient than the 3-point seed method. This algorithm has been applied to the object of Fig. 8, yielding an approximation with triangles. The edges of these triangles are visible on the Figure.

Nonplanar approximations

Many man made objects and in particular industrial parts cannot be represented very conveniently with planar primitives only. Therefore we have added to these planar primitives simple non-planar ones, i.e. cones and cylinders because they are useful for object modelling and recognition and can be recovered from range data fairly easily.

This is done by using the normal to the surface of the object at a measured point. This vector can be easily computed from the range data and is represented (after normalisation) as points on the unit sphere. Suppose first that a part of the object is planar. All the measured points within that part have approximately the same normals and we find a tight cluster of corresponding points on the unit sphere. If we have a way of detecting such tight clusters we are capable of detecting the planar parts of the object. Suppose then that a part of the object is cylindrical (conical). All the normals will fall on a large circle (a circle) of the unit sphere. If we have a way of detecting such circular clusters, we are capable of detecting the cylindrical and conical parts of the object.

It turns out that this cluster detection can be efficiently implemented using a technique known as the Hough transform. Of course it does not allow us to discriminate between distinct but parallel planar faces or between cylinders or cones of parallel axis. This discrimination is achieved by going back to the original set of data points and using notions of connectivity. The algorithm has the following steps.

1. Find the set of planar faces and remove the corresponding points from the list of data points.
2. Find the set of cylindrical points and remove them from the list of data points.
3. Find the set of conical points and remove them from the list of data points.

Steps 1, 2 and 3 are hierarchically structured in the sense that we first look for the largest planar faces, cylinders, cones and then for smaller ones. At each step we also separate distinct primitives of the same class like parallel cylinders or planar faces.

The Renault part which is shown in Fig. 2 is actually mostly made of planes and cylinders. We have applied this technique[23] to the 0° view of Fig. 3 and the results are shown in Fig. 9.

Volumic representations

An alternative approach is to use elementary volumes to describe objects.

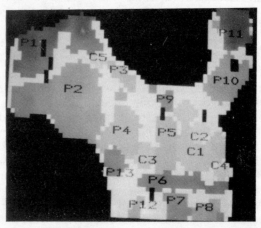

Fig. 9. Planes and cylinders found in the view of the Fig. 3 by the Hough Transform. Planes are labelled from P1 to P13 and cylinders from C1 to C5.

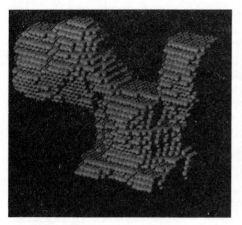

Fig. 10. The Decomposition of the part of the Fig. 2 in rhombododecahedra (a), and the skeleton obtained (b).

Although expensive in terms of memory, these methods have several advantages, such as making easy the computation of integral properties. Our approach is to use the rhombododecahedron as unit volume. This polyhedron, as described in,[24] is closer to the sphere, and more isotropic, than the cube (as, in the planar case, the hexagon is 'better' than the square). This makes it a good tool for 3-D skeletonisation. Moreover, the object decomposition into rhombododecahedra is quite easy from equidistant slices obtained from the laser range finder. Fig. 10 shows the decomposition of the automobile part of Fig. 2, and the skeleton obtained. We look forward to extend this work to the construction of hierarchical structures, the 13-Trees, which are analoguous to the Octrees of the cubic grid.

3-D object recognition

Our approach for 3-D shape matching so far uses planar faces as primitives and matches an unknown view with the structural 3-D model. Shape matching is therefore performed by matching the face description of an unknown view with the stored model. The descriptors of a face in both the model and an unknown view are:
- ○ the area, i.e. the number of points belonging to the face,
- ○ the perimeter, i.e. the number of points on the boundary,
- ○ the length of the maximum, minimum and average radius vectors from the centroid,
- ○ the number of vertices in the polygonal approximation of the boundary,
- ○ the angle between the maximum and minimum radius vectors,
- ○ the ratio of area/perimeter^2,
- ○ the coordinates of the centroid,
- ○ the equation of the plane containing the face,
- ○ its neighbours.

Matching algorithm

Our task is to match a model against an observed object. The model and the object are represented as sets of geometric primitives $T = (T_1, T_2, \ldots, T_N)$ and $O = (O_1, O_2, \ldots, O_{L-1})$, respectively. These primitives are planar faces. They are structured spatially in the following manner. We consider binary relations between primitives such as adjacency, parallelism, etc. . . . This induces a labelled graph structure on the sets T and O where an arc between T_i AND T_j are adjacent or parallel, etc.

Nodes of the graphs are also labelled with feature values such as length, area, angle with adjacent primitives, etc.

Model elements are referred to as units and object elements as labels. The problem of matching can thus be expressed as that of labelling each of the units T_i (i = 1,2, . . . , N) either as a label O_j (j = 1,2, . . . , L-1) or as not belonging to O (label O_L = NIL). Each unit T_i has therefore L possible labels. Another equivalent way of thinking about the matching problem is to see it as a problem of matching two labelled graph structures. This problem has recently received much attention[17-21] and the solution that we propose here as in[21] is based on finding maxima of a simple criterion.[22]

Using a technique described below, we compute for every face T_i a set of L positive numbers p_i (l), l = 1,2, . . . , L. p_i (l) can be thought of as the probability of labelling the face T_j as O_l. The initial probabilities are computed using the first six features in the above list. We measure the quality of the correspondence between the faces T_i and O_k as:

$$M(T_i, O_k) = \sum_{i=1}^{P} |f_{tp} - f_{op}| W_p$$

where, f_{tp} = pth feature value for the face of an unknown view,
         f_{op} = pth feature value for the face of the model,
         w_p = weight factor for the pth feature.

We need the weights of the features to account for their importance and range of values. Also any a priori knowledge that we may have regarding the type of change or deformation can be incorporated into these weights. The initial probabilities are chosen proportional to $1/(1+M(T_i, O_k))$ and normalised so that they sum to 1. We also introduce a compatibility function $c(T_i, T_j, O_k, O_l)$ to determine the degree by which the assignments of two neighbouring units T_i and T_j are compatible with each other. The key idea in their computation is that of spatial transformation. In order to compute $c(T_i, T_j, O_k, O_l)$ (for k and *l* not equal to L) we define two transformations:

$$TR1 : T_i \rightarrow O_k \text{ and,}$$

$$TR2 : T_j \rightarrow O_l$$

The transformations are based on:
○ scale, the ratio of the areas of two faces,
○ translation, difference in the centroidal coordinates of the two faces,
○ orientation, difference in the orientation of the two faces so that they are in the same plane,
○ rotation, to obtain maximum area of intercept, once the two faces are in the same plane.

TR1 is then applied to T_j and TR2 to T_i and the resulting errors Error (TR1(T_j),O_l) and Error (TR2(T_i),O_k) are computed. The compatibility value $c(T_i, T_j, O_k, O_l)$ is set equal to $1/(1+\text{Average Error})$. Just as the p_i (l)'s represent what the face T_i 'thinks' about its own labelling, the function c is used to build numbers q_i (l) that represent what the neighbouring faces of T_i 'think' about its labelling. Because of imperfect measurements P_i (l) and q_i (l) tend to disagree in general in the beginning.

The purpose of our algorithm is to bring them into agreement. This can be done efficiently by maximising a simple criterion, a function of the p_i (l)'s. The initial value of the criterion is computed from the initial probabilities and a local maximum is sought from there. The maximum yields a labelling of the faces T_i which is better than the original one. This labelling provides in turn the geometrical transformation from the model to the observed object.

Examples

We present examples using the industrial object shown in Fig. 2. The number of faces in an unknown view varied from 10 to 26 and the number of faces in the model is 85. In the matching process, only the best 29 faces of the model are considered in order to reduce the complexity. In testing the shape matching algorithm we consider three unknown views corresponding to 0°, 30° and 330° respectively. Although the model is obtained using these views, it does not contain the faces corresponding to any unknown view. This is because of the nature of how the surface points corresponding to the composite object were obtained.

The results of the labelling are used to compute the relative orientation of an unknown view with respect to the model. The method requires the knowledge of the neighbours of a face. These neighbours are arranged in descending order according to their size and the larger neighbours are preferred to the smaller ones. The results of the labelling are used to compute the relative orientation of the unknown view with respect to the model. In order to do this, we need at least three matched pairs of faces. Since some units may be in error we in fact compute an average of several solutions obtained from different triplets of matched faces.

The matrices of the direction cosines of a coordinate system tied to the object relative to the coordinate system tied to the model are found to be:

computed			theoretical		
1.0	0.0	0.0	1.0	0.0	0.0
0.0	1.0	0.0	0.0	1.0	0.0
0.0	0.0	1.0	0.0	0.0	1.0
0.9	0.1	0.5	0.9	0.0	0.5
−0.2	1.0	−0.2	0.0	1.0	0.0
−0.4	0.0	0.9	−0.5	0	0.9
0.7	0.0	−0.6	0.9	0.0	−0.5
−0.1	1.0	0.0	0.0	1.0	0.0
0.5	0.0	0.7	0.5	0.0	0.9

The computed matrices are on the left, the theoretical ones on the right. Thus the matching results provide reasonable orientation information.

Conclusion

We have presented a computer controlled system for reliably and accurately acquiring 3-D range data. This system enables us to produce automatically geometric models of objects which can be used to compute properties of these objects such as stable positions, graspable configurations, etc., and to guide a vision system for recognition of objects.

The geometric models used so far are fairly simple and we hope to refine them in the future and articulate them with CAD produced models. The recognition of objects has been implemented solely for isolated objects and using only range data. We plan to extend our work in two directions: the co-operation of a TV camera measuring light intensities and the laser rangefinder measuring distances and the recognition of partially occluded objects.

References

[1] J. D. Boissonnat and F. Germain, 'A New Approcah to the Problem of Acquiring Randomly Oriented Workpieces out of a Bin', *Proc. 7th Int. Joint Conf. On Artificial Intelligence*, pp. 796-802 (1981).

[2] T. O. Binford, 'Visual Perception by Computer', *IEEE Conf. on Systems and Control*, Miami (December 1971).

[3] G. J. Agin and T. O. Binford, 'Computer Description of Curved Objects', *Proc. 3rd Int. Joint Conf. on Artificial Intelligence*, pp. 629-640.

[4] R. Nevatia and T. O. Binford, 'Description and Recognition of Curved Objects', *Artificial Intelligence*, Vol. 8, pp. 77-98 (1977).

[5] R. K. Bajcsy and B. I. Soroka, 'Steps towards the Representation of Complex Three-Dimensional Objects', *Proc. 5th Int. Joint Conf. on Artificial Intelligence*, pp. 596.

[6] D. Marr, 'Representing Visual Information – A Computational Approach', in *Computer Vision Systems*, A. R. Hanson and E. M. Riseman (Eds.), pp. 61-80, Academic Press (1978).

[7] N. Badler and R. Bajcsy, 'Three-Dimensional Representation for Computer Graphics and Computer Vision', *ACM Computer Graphics*, Vol. 12, pp. 153-160 (August 1978).

[8] M. Oshima and Y. Shirai, 'A Scene Description Method Using Three-Dimensional Information', *Pattern Recognition*, Vol. 11, pp. 9-17 (1978).

[9] Y. Shirai, 'On Application of 3-Dimensional Computer Vision', *Bul. Electrotech. Lab.*, Vol. 43, No. 6 (1979).

[10] Y. Shirai, 'Use of Models in Three-Dimensional Object Recognition', in *Man-Machine Communication in CAD/CAM*, T. Sata, E. Warman (Eds.), North-Holland Publishing Company (1981).

[11] B. Bhanu, 'Shape Matching and Image Segmentation Using Stochastic Labelling', Ph.D Dissertation, University of Southern California, Los Angeles, August 1981.

[12] D. T. Lee and B. J. Schacter, 'Two Algorithms for Constructing a Delaunay Triangulation, *Int. Journal of Comp. And Inf. Sciences*, Vol. 9, No. 3 (1980).

[13] F. P. Preparata and S. J. Hong, 'Convex Hulls of Finite Sets of Points in Two and Three Dimensions', *Comm. ACM*, Vol. 20, No. 2 (1977).

[14] D. Hilbert and S. Cohn Vossen, *Geometry and Imagination*, Chelsea.

[15] J. L. Bentley, J. H. Friedman, 'Fast Algorithms for Constructing Minimal Spanning Trees in Coordinate Spaces', *IEEE Trans. Comp.*, Vol. C-27, No. 2 (February 1978).

[16] J. D. Boissonnat and O. D. Faugeras, 'Triangulation of 3-D Óbjects', *Proc. 7th Int. Joint Conf. on Artificial Intelligence*, pp. 658-660 (1981).

[17] L. Kitchen and A. Rosenfeld, 'Discrete Relaxation for Matching Relational Structures', *IEEE Trans. Systems, Man and Cybernetics*, Vol. SMC-9, pp. 869-874 (Dec. 1979).

[18] L. Kitchen, 'Relaxation Applied to Matching Quantitative Relational Structures', *IEEE Trans. Systems, Man and Cybernetics*, Vol. SMC-10, pp. 96-101 (February 1980).

[19] R. M. Haralick and L. G. Shapiro, 'The Consistent Labelling Problem, Part I', *IEEE Trans. Pattern Anal. Machine Intell.*, Vol. PAMI-I, pp. 173-184 (April 1979).

[20] R. M. Haralick and L. G Shapiro, 'The Consistent Labelling Problem, Part II', *IEEE Trans. Pattern Anal. Machine Intell.*, Vol. PAMI-2, pp. 193-203 (May 1980).

[21] O. D. Faugeras and K. Price, 'Semantic Description of Aerial Images Using Stochastic Labelling', *IEEE Trans, Pattern Anal. Machine Intell.*, Vol. PAMI-, pp. (November 1981).

[22] O. D. Faugera and M. Berthod, 'Scene Labelling: an Optimization Approach', *IEEE Trans. Pattern Anal. Machine Intell.*, Vol. PAMI-, pp. (1981).

[23] M. Hebert and J. Ponce, 'A New Method For Segmenting 3-D Scenes'. *I.C.P.R.* (October 1982).

[24] J. Serra, *Image Analysis and Mathematical Morphology*, Academic Press (1981).

[25] E. Pauchon, These de Docteur Ingenieur, Universite d'Orsay, 1983, to appear.

[26] N. Ayache, 'Reconnaissance d'objets 2-D parteillement caches', These de Docteur Ingenieur, Universite d'Orsay, to appear December 1982.

THE USE OF TAPER LIGHT BEAM FOR OBJECT RECOGNITION

Dai Wei* and M. Gini, Department of Electronics, Polytechnic of Milano, Italy

Abstract

A recognition procedure with a taper light beam is presented. The method employs a vertical projector which projects a taper light beam on the objects where it produces an elliptical or circular bright area. A picture is then taken by a TV camera. Features of the objects such as the distance to each point, and the angle between the surfaces and the light beam can be obtained by means of trigonometrical computations. The recognition procedure is free from the effects of the arrangement and shadows of objects, and the measurement of parameters concerned is more precise. The mathematical aspects of the method are given.

Introduction

The recognition of three-dimensional objects is an important part of machine vision.[1] In fact it plays an important role in robotics, where there is a recurrent need to recognise the environment.

Different methods have been proposed for the recognition of 3D objects.[2] One of the most complete is based on the use of stereo vision[3] which allows complete models to be constructed at a fairly high computational cost. Other approaches have appeared. For instance Shirai[4] designed a method based on the use of a vertical slit projector and a TV camera, to pick up the reflected light. By rotating the projector from left to right the distance of many points in the field of view can be computed. Other authors have proposed solutions based on the use of laser[5] or of planes of light that are projected on the objects to determine their shape.[6]

The problem with most of those methods is that they tend to be expensive from the point of view of equipment and/or computation time, so that their use in industrial environments is seldom justifiable.

Simpler methods based on the use of two dimensional images are often proposed for industrial applications.[7] One problem with methods based on 2D images is that if two different objects have the same projection on the 2D plane, as shown in Fig. 1, it is impossible to distinguish them. In these cases the heights and the angles of the polyhedrons, which are very important to recognise the objects, cannot be correctly determined.

In this paper we propose a method called 'taper light beam projection'. By this method we can obtain more knowledge about the objects, more precisely, and at a

---

*Dai Wei is a Chinese engineer engaged in research as visiting scholar at the Department of Electronics of the Milan Polytechnic.

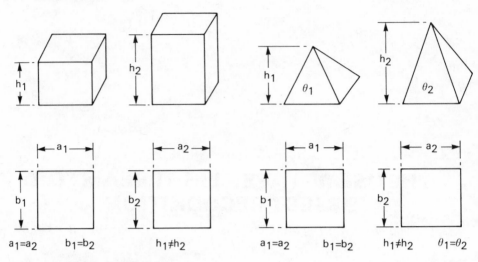

$a_1 = a_2$ $b_1 = b_2$ $h_1 \neq h_2$ $a_1 = a_2$ $b_1 = b_2$ $h_1 \neq h_2$ $\theta_1 = \theta_2$

Fig. 1. Different objects that can be seen as equal.

reduced cost. Our method applies well to those applications in which the 2D information extracted by the silhouettes of the objects is not sufficient although a complete 3D analysis is too expensive.

The method of taper light beam projection

As shown in Fig. 2, let A be a TV camera which is over the object K. From the camera a beam of light is projected vertically downwards on the object K. This taper light beam will form an elliptical or circular bright line, say G, on a certain surface of the object, say M. From the shape of this bright line we can compute the distance between the camera and the centre of the bright line, the direction of slope, and the angle between the axis of the taper light beam and the plane on which the bright line is projected.

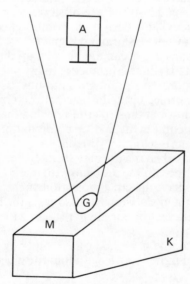

Fig. 2. The arrangement of camera and taper light beam.

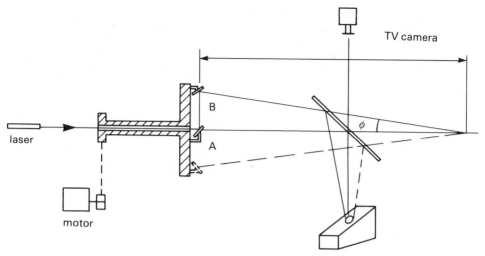

Fig. 3. The experimental set-up of the system.

The experimental set-up of the system, which is now being developed, is based on the use of a laser for the generation of the light beam, illustrated in Fig. 3.

We have a Ne-He laser tube, which gives a thin and strong light beam, and a turning plate, which can change the straight laser beam into a taper light beam when it is moved by the motor. The mirror A is used to change the direction of the light beam, the mirror B can be adjusted to change the angle and the distance between the object and the camera. A translucent mirror can change the direction of the taper beam so that the TV camera can take the picture of the line of light formed on the surface of the object under consideration.

If we move the camera together with the source of light on a plane parallel to the working plane (or, alternatively we move the object under the camera) we may determine the shape of the object in a more complete way. In our system the object is on a moving table so that it will be moved on a plane under the source of light in any direction. The number of points in which the method is applied and their positions depends on various choices which can be left to the user responsibility or can be resonably automated according to the objects that we want to inspect.

For simplicity let us suppose in this presentation to have both the TV camera and the projector of light at the same point.

Mathematical analysis of the method

Suppose that, as shown in Fig. 4, A is a camera. At the same point A, there is also a taper light beam projector, which focuses at the point S. Half of the cone angle of this taper light beam is ϕ. The distance between S and A is L_0. The ellipse BECF is formed by the beam on the top surface of the object.

Let BC be the longer axis of the elliptical bright area, ψ_1 and ψ_2 respectively be the angles between the axis AS and the lines AC and AB, D the middle point of BC, L_1 the distance between the camera and the point D which is the centre of the ellipse.

We have the following equations:

$$L_2 \cdot \operatorname{tg} \psi_1 = (L_0 - L_2) \operatorname{tg} \phi \tag{1}$$

$$L_3 \cdot \operatorname{tg} \psi_2 = (L_0 - L_3) \operatorname{tg} \phi \tag{2}$$

$$\overline{BC}^2 - (L_3 - L_2)^2 = ((L_0 - L_3)\, \text{tg}\phi + (L_0 - L_2)\, \text{tg}\phi))^2 \tag{3}$$

$$L_1 = L_2 + (L_3 - L_2)/2 \tag{4}$$

$$\text{ctg}\alpha = \frac{L_3 - L_2}{(L_0 - L_3)\,\text{tg}\phi + (L_0 - L_2)\,\text{tg}\phi} \tag{5}$$

$$\overline{OR} = (L_0 - L_1)\,\text{tg}\phi \tag{6}$$

$$\overline{OD} = (\overline{BC}/2)\sin\alpha + (\overline{BC}/2)\cos\alpha\,\text{tg}\phi - \overline{OR} \tag{7}$$

$$(\overline{EF}/2)^2 = \overline{OR}^2 - \overline{OD}^2 \tag{8}$$

From these equations we can compute:

$$\text{ctg}\alpha = \frac{L_0 \cdot \text{tg}\phi\,(\text{tg}\psi_1 - \text{tg}\psi_2)}{(\text{tg}\psi_1 + \text{tg}\phi)(\text{tg}\psi_2 + \text{tg}\phi)} \tag{9}$$

$$L_1 = \frac{L_0\,\text{tg}\phi\,(\text{tg}\psi_1 + \text{tg}\psi_2 + 2\,\text{tg}\phi)}{2(\text{tg}\psi_1 + \text{tg}\phi)\,(\text{tg}\psi_2 + \text{tg}\phi)} \tag{10}$$

$$L_2 = \frac{L_0\,\text{tg}\phi}{\text{tg}\psi_1 + \text{tg}\phi} \tag{11}$$

$$L_3 = \frac{L_0\,\text{tg}\phi}{\text{tg}\psi_2 + \text{tg}\phi} \tag{12}$$

$$\overline{BC} = L_0 \sqrt{\text{tg}^2\phi\left[\frac{\text{tg}\psi_1 - \text{tg}\psi_2}{\text{tg}\psi_1 + \text{tg}\psi)(\text{tg}\psi_2 + \text{tg}\phi)}\right]^2 + \left[\frac{\text{tg}\psi_1}{\text{tg}\psi_1 + \text{tg}\phi} + \frac{\text{tg}\psi_2}{\text{tg}\psi_2 + \text{tg}\phi}\right]^2} \tag{13}$$

$$\overline{EF} = 2\sqrt{\overline{BC}(L_0 - L_1)(\sin\alpha + \cos\alpha\,\text{tg}\phi)\,\text{tg}\phi - \overline{BC}^2(\sin\alpha + \cos\alpha\,\text{tg}\phi)^2} \tag{14}$$

The angle ϕ and the distance L_0 are known; they depend on the characteristics of the projector. The angles ψ_1 and ψ_2 are parameters which we can determine according to the position of the bright elliptical area in the field of view. BC and EF are respectively the longer and shorter axes of the ellipse.

The most important parameters that we may compute are α and L_1, which represent respectively the angle of the top surface of the object and its distance from the camera.

Computation of parameters from the image

From the above we know that the parameters ψ_1 and ψ_2 are important for the information that we want. They are the only parameters which we have to determine from the image.

We find that ψ_1 and ψ_2 are the biggest and the smallest of all angles between every point on the edge of the bright elliptical line and the line AS. We give the deduction as follows.

In Fig. 5, A is the camera whose focal axis coincides with the axis of the taper light

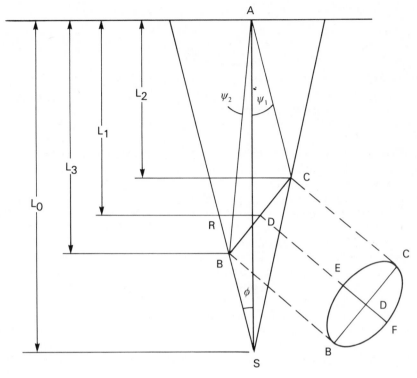

Fig. 4. The proposed method.

beam, say AS. K is an arbitrary point on the edge of the ellipse. Let R be the distance between the point K and the axis AS. The points B and C are respectively the lowest and the highest points of the ellipse; let L_b, L_c and L_k be respectively the distance from the points B, C, and K to the plane on which the projector is positioned.

We have the following equations:

$$R = (L_0 - L_k)\,\text{tg}\,\phi \tag{15}$$

$$\text{tg}\,\psi_k = R/L_k \tag{16}$$

The angle ψ_k is the angle between AK and AS. From the equations (15) and (16) we can get:

$$\text{tg}\,\psi_k = (L_0/L_k - 1)\,\text{tg}\,\phi \tag{17}$$

According to this formula $\text{tg}\,\psi_k$ will increase when L_k decreases, and vice versa.

We know that the two terminal points of the longer axis of the ellipse are the highest one, the point C, and the lowest one, the point B, on the bright elliptical line. Hence L_B and L_C are the smallest and biggest values of the distance. From equation (17) we can conclude that the angles ψ_C and ψ_B are the largest and smallest of angles between every point on the ellipse and the axis AS.

Obviously if the top surface of the object is parallel to the working plane, ψ_1 and ψ_2 and all the ψ_k angles are equal.

Knowing the focal length F of the camera we can compute from the image:

$$\text{tg}\psi_k = \frac{D'K'}{F}$$

(18)

where $D'K'$ is the projection on the same image of the segment DK.

According to the previous considerations to determine ψ_1 and ψ_2 we have to find out on the image the biggest and smallest values of ψ_k.

How to determine different surfaces of a polyhedron

Consider the object shown in Fig. 6a. By moving the camera we take two different images of the light projected on it.

If we have computed two different angles α_1 and α_2 by equation (9) at two different positions, we can surely conclude that these parts are on two different surfaces. We will show later on (see equation (23)) how to compute the angle θ between the surfaces of the polyhedron.

Consider now the situation shown in Fig. 6b. If with different applications of the method we obtain the same value of the angle α, as shown in Fig. 6, the question of determining the surfaces of the object becomes complex. The value of α does not allow a judgement to be made on whether the different points are on the same surface, such as points A and B, or not, such as B and C.

In this case we have some more considerations. Let us look more precisely at Fig. 7.

Let L_{AB} be the distance between the two camera positions when applying the

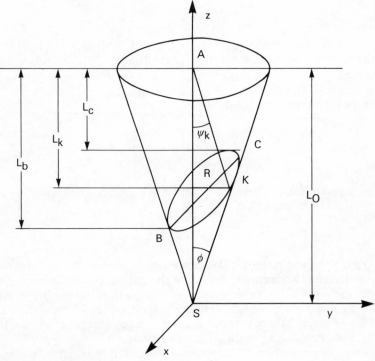

Fig. 5. How to compute the angles.

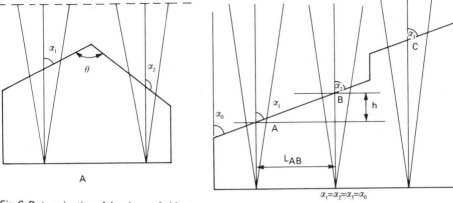

Fig. 6. Determination of the shape of objects.

$\alpha_1 = \alpha_2 = \alpha_3 = \alpha_0$

B

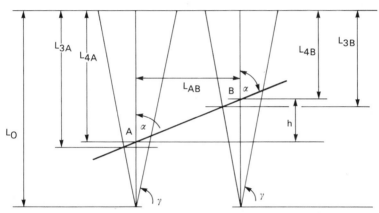

Fig. 7. Determination of the top surface of an object.

method; L_{3A} is the value of L_3 computed by equation (12) at point A, L_{3B} the value of L_3 at point B. We can compute:

$$L_{4A} = L_{3A} - (L_0 - L_{3A})\,\mathrm{tg}\phi\,\mathrm{ctg}\alpha \tag{19}$$

$$L_{4B} = L_{3B} - (L_0 - L_{3B})\,\mathrm{tg}\phi\,\mathrm{ctg}\alpha \tag{20}$$

$$h = L_{4A} - L_{4B} = (L_{3A} - L_{3B})\,(1 + \mathrm{tg}\phi\,\mathrm{ctg}\alpha) \tag{21}$$

$$\mathrm{ctg}\alpha = h/L_{AB} \tag{22}$$

If equation (22) is true, A and B are on the same surface, otherwise they are on different surfaces.

Measurement of the angle between two adjacent surfaces

The angle between two adjacent surfaces is an important parameter for the recognition of a polyhedron. If we have the situation shown in Fig. 8, firstly we compute the angles ψ_1 and ψ_2 by means of the equation (18). Thus we know the direction of the longer axes of the ellipses.

If on two different planes, say M and N, we get two different directions of the two longer axes of the ellipses, β_1 and β_2, we can compute the angle θ, which is the angle between the plane N and M, by:

$$\cos \theta = \sin \alpha_1 \sin \alpha_2 + \cos \alpha_1 \cos \alpha_2 \cos (\beta_1 - \beta_2) \qquad (23)$$

where α_1 and α_2 are respectively the angle α in equation (9) on plane M and plane N (see Fig. 3). This formula can be obtained by the following considerations. Let $\vec{\delta_1}$ be the unit vector perpendicular to the surface M, and $\vec{\delta_2}$ the same for the surface N. We have:

$$\vec{\delta_1} = (\cos \alpha_1 \cos \beta_1)\,\vec{i_1} + (\cos \alpha_1 \sin \beta_1)\vec{j} + \sin \alpha_1 \vec{\kappa} \qquad (24)$$

$$\vec{\delta_2} = (\cos \alpha_2 \cos \beta_2)\,\vec{i_1} + (\cos \alpha_2 \sin \beta_2)\vec{j} + \sin \alpha_2 \vec{\kappa} \qquad (25)$$

$$\cos \theta = \frac{\vec{\delta_1} \cdot \vec{\delta_2}}{|\vec{\delta_1}| \cdot |\vec{\delta_2}|}$$

$$= \frac{(\cos \alpha_1 \cos \beta_1)\ (\cos \alpha_2 \cos \beta_2)\ +}{1 \cdot 1}$$

$$\frac{(\cos \alpha_1 \sin \beta_1)(\cos \alpha_2 \sin \beta_2) + \sin \alpha_1 \sin \alpha_2}{1 \cdot 1}$$

$$= \cos \alpha_1 \cos \alpha_2 \cos (\beta_1 - \beta_2) + \sin \alpha_1 \sin \alpha_2 \qquad (26)$$

Finding the edges

We show in Fig. 9 what can happen in the case where the beam of light is projected on one edge of a polyhedron.

Suppose that $\vec{\delta_R}$, $\vec{\delta_P}$, $\vec{\delta_Q}$, and $\vec{\delta_N}$ are the unit vectors which are respectively perpendicular to the surfaces R, P, Q, and N. Let θ_R, θ_P, θ_Q, and θ_N be the angles between the direction of the taper light beam and the unit vectors. We may be in one of two different situations. Both the angles of the surfaces which form the edge

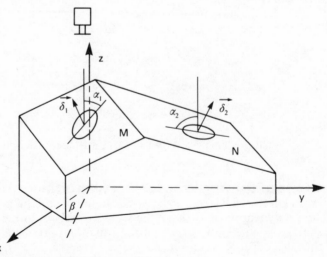

Fig. 8. Computation of angles between two adjacent surfaces.

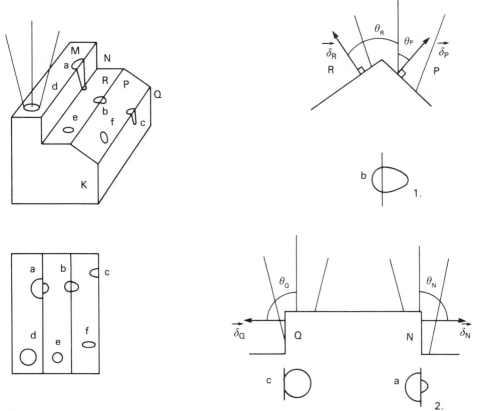

Fig. 9. Different edges of a polyhedron.

area are less than 90 degrees as in the case 1 of Fig. 9, or one of the angles is equal to or larger than 90 degrees as in the case 2.

If we project the light on the edges of the polyhedron K we can see different forms of bright rings as A, B, or C. A and C are produced on the edges where the angle θ_Q or θ_N is equal to or larger than 90 degrees while B is produced on the edge where the angles θ_R and θ_P are less than 90 degrees. Comparing those bright rings with D, E and F, which are only on one plane, we can find that they are quite different.

Let us now consider the cases 1 and 2 of Fig. 9. In this case 2 (θ_N greater or equal to 90 degrees) we can easily find the position on the edge. In fact during the movement of the object if we find a closed line which becomes open we can surely conclude that at this moment we have discovered one edge. However in the case 1 the problem becomes more complex because the form B is similar to D, E, and F. In this case we can find a difference between them by differential calculus. In the cases D, E and F the differentials of the ring equations are continual but in the case B the differential of the ring is not continual.

Applications of the method of taper light beam

Let us now consider an example of the application of a combination of vision with the taper light beam projection. In Fig. 10a there are three objects. A is a line drawing, B and C are real objects.

If a TV camera is over them the pictures taken for the three different objects are almost the same. But if we use the taper light beam on every visible surface of the objects we will find that in the case A the distances and the angles are the same for all the five points considered, while in the other cases we obtain different results.

One important aspect is how to decide where to apply the method. In this case vision can help, allowing us to determine how many visible surfaces the object has and where they are, so that the taper light beam can be projected over them. In this case there are five visible surfaces and so the method is applied exactly five times.

A more general approach is based on a regular movement of the taper light beam over the object to determine a regular grid of points on it. At each point a picture is taken and the values of α and L_1 are computed.

Let us see another example. In Fig. 10b we have three objects, which differ only in the top surface: in object A the top surface is flat, while in B it is convex and in C it is concave. The images taken by a TV camera over them are similar, but the use of the taper light beam can determine the difference. If the source of light and the

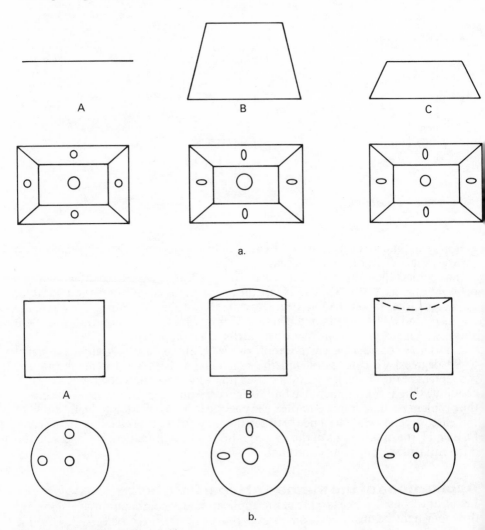

Fig. 10. Recognition of different objects.

camera are moved in a regular way over the object we can recover more detailed information. In fact the distance of the object from the camera is the same only in case A.

Conclusions

We have presented a method for recognition of features of three dimensional objects based on the idea of projecting on them a taper light beam. The method is fairly accurate and computationally simple.

Acknowledgements

The reported work is partially financed by Progetto Finalizzato per l'Informatica, Obiettivo MODIAC. The authors would like to thank M. Somalvico for his kind guidance. Thanks are also due to R. Cassinis for advice and helpful comments.

References

[1] H. P. Winston, 'The psychology of computer vision', McGraw Hill, 1975.
[2] G. G. Dodd and L. Rossol, 'Computer vision and sensor-based robots', Plenum Press, New York, 1979.
[3] M. Bernasconi, R. Delazer and M. Gini, 'Accuracy of measurement through stereo images', Proc. Automated Inspection and Product Control, Stuttgart, Germany (June 1980).
[4] Y. Shirai and M. Suwa, 'Recognition of polyhedrons with a range finder', Proc. IJCAI-71, London, UK (1971).
[5] G. Agin and T. O. Binford, 'Computer description of curved objects', Proc. IJCAI-73, Stanford, Ca. (August 1973).
[6] R. Popplestone, *et. al.*, 'Forming models of plane-and-cylinder faceted bodies from light stripes', Proc. IJCAI-75, Tbilisi, URSS (September 1975).
[7] G. Gini and M. Gini, 'A general purpose vision sensor', Proc. 26th SPIE, San Diego, Ca. (August 1982).

Chapter 4
DEVELOPMENTS –
WELD GUIDANCE

Problems of part fit-up and jigging are inhibiting the wide spread application of robots to arc welding. Vision guidance of the robot offers the best promise of solving this problem. Many of the commercial companies offering robot welding systems are developing vision techniques for robot path control. Three of the papers describe such systems which are now available and the fourth paper describes industrially sponsored research on the subject.

PRESENT INDUSTRIAL USE OF VISION SENSORS FOR ROBOT GUIDANCE

Philippe Villers, Automatix Inc., United States of America

Presented at 12th International Symposium on Industrial Robots – Paris 9-11 June, 1982

Abstract

Intelligent robot systems are "intelligent" because of their ability to adapt to their environment. This implies powerful real time computer systems and advanced sensors such as vision. In this paper we discuss the practical application of vision to 100% part inspection, robot guidance for locating and taking up parts, and the application of vision to provide servo position feedback for arc welding. In the case of arc welding, departures from the required part to part nominal path as well as information concerning the amount of weld metal required to fill a gap is provided by the vision system. Industrial examples using Autovision, as exemplified in the Cybervision System and Robovision II Systems, will be discussed and their significance for certain classes of industrial use explained.

Introduction

The critical role of artificial vision in increasing the range of robot applications has been understood for some time, as evidenced by the intensive research efforts waged on three continents. But this is the year, 1982, when vision systems will leave the laboratory. Commercial systems for both assembly and arc welding demonstrated in March at the Robots VI conference in Detroit, Michigan, U.S.A., are being put to work on the factory floor. This long awaited event will undoubtedly have a major impact on the kind and range of tasks routinely performed by robots.

This paper examines one of the first such commercial systems, the Automatix Autovision II, the only commercially produced vision unit known to the author to be in commercial use for robot seam tracking. It explores principles of operation and provides examples of present applications in automotive metal stamping, injection molding and PC board assembly to show the versatility of use of modern micro-processor based artificial vision systems, as they presently work in industry.

Classes of artificial vision applications

Artificial vision systems for robot guidance take basically two forms. One is verification of parts or position or determination for assembly, material handling, and related tasks. In such an application the vision system may allow the robot to pick a part off a moving belt, as in General Motors' *CONSIGHT* system, or for

Fig. 1. Autovision II system.

assembling parts where initial part position and orientation is not known sufficiently accurately. The vision system may also be used for verification and feedback of completion of assembly operations – such as in the Westinghouse *APAS* batch motor assembly, the Automatix Cybervision System or Hitachi's experimental unfixtured vacuum cleaner final assembly. In all of these, the artificial vision system is based on real time computer interpretation of television images using available or specialised (structured) lighting. These tasks mainly involve recognition of the part, computation of its centre of gravity and orientation, and the passing of this information in real time to a robot.

The second class of robotic vision guidance task is visual servo control such as in seam tracking for arc welding. This is a fundamentally different task: using vision information to provide position information to the servo control of the robot to adjust its path.

Despite the difference in the two classes of application, both can use the same television camera and micro-processor based analysis system with appropriate modifications. In the first class of applications the camera is more likely to be in a fixed location; in the second to be held by the robot itself.

The vision sub-system

The Autovision II system as shown in Fig. 1 is a commercial descendant of the earlier research work done at Stanford Research Institute in California, using a fixed position vision system, and the "camera in the hand" work done at the National Bureau of Standards using robot mounted vision. It retains from its *SRI* origins the data compaction using run length encoding as described in[1], and the powerful concept of feature extraction as described in [2] which reduces the task of identifying an object and its position to the computation of suitably chosen (typically a half dozen) extracted geometric features from an available list of approximately 50. These include such basic parameters as area, perimeter, number of holes, major and minor axes of equivalent ellipse, moment of inertia, etc. Thus the "feature extraction" approach reduces the comparison or "recognition" of pictures or patterns to the comparison of a small series of numbers, each describing a feature. Most tasks are accomplished using binary vision for speed and simplicity but a gray scale capability for those applications which require gray scale has recently been added.

In order to gain high speed performance, (in one automotive inspection application, the Autovision II is used to inspect 2,200 holes of different sizes and shapes per minute) a high speed bi-polar processor is used for pre-processing followed by a powerful 8 mHz 32 bit arithmetic micro-processor, the Motorola 68000, with its ability to address a million word memory. As a commercial product, the Autovision II is housed in a sealed environment with cooling provided by a heat exchanger to ensure operation in adverse factory environments such as an arc welding shop or foundry.

A block diagram of the Autovision II is shown in Fig. 2. The Autovision system [3] shown is used not only for robot guidance, but for 100% real time high speed visual inspection ranging from large automobile stampings coming off the press every three seconds, such as those of Fig. 3, to the 7 mm injection moulded plastic parts shown in Fig. 4, which are carried past the camera in random orientation and position for 100% inspection at 360 per minute.

High contrast backlighting of the part is used where feasible. However, often backlighting cannot be used and front lighting is necessary, as in the use of uniform

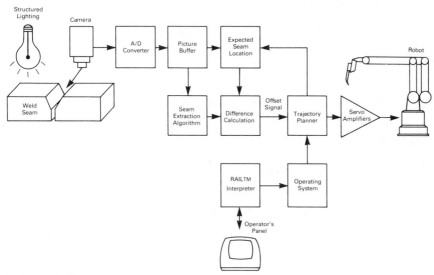

Fig. 2. Autovision II Block Diagram.

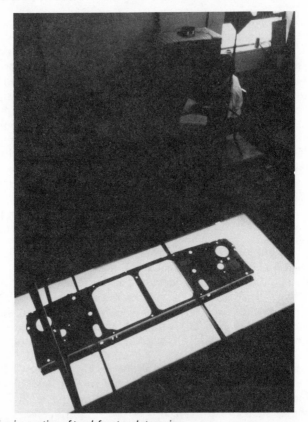

Fig. 3. 100% vision inspection of truck front end stamping.

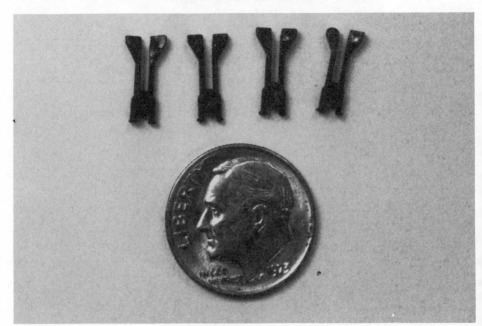

Fig. 4. High speed inspection of injection moulded parts.

white light for verification of parts to be assembled and assembly position guidance, or the use of laser light in the case of seam tracking for robot arc welding.

Programming vision systems

Teaching advanced vision systems to recognise a new part typically requires less than one hour's work and involves aiming the cameras so the part or the to-be-inspected portions of large parts can be seen by at least one camera regardless of the position of the part when it arrives. As each part or selected segment is automatically displayed on the screen during teaching, the operator is asked to answer questions about permissible tolerances or criteria and selected features using a simple computer language called *RAIL* (Robot Automatix Inc. Language). Training for recognition of a new part normally requires 10 or more samples to be presented. The vision system extracts geometric features and for each calculates mean and standard deviation as an aid in user feature selection.

The simplicity of *RAIL* language commands is illustrated in Fig. 5. Here the problem is to check for correct mounting of a hybrid chip on a heat sink while moving down a conveyor belt. The permissible tolerances are expressed as maximum permissible x error, y error, and θ error.[4]. The *RAIL* program to perform this task is as follows:

```
INPUT PORT 1: CONVEYOR
 INPUT PORT 2: PART_DETECTOR
 OUTPUT PORT 1: BAD_PART

 WRITE "ENTER CHIP OFFSET TOLERANCE:"
 READ OFFSET_TOL
 WRITE "ENTER CHIP TILT TOLERANCE:"
 READ TILT_TOL

 WAIT UNTIL CONVEYOR = ON
 WHILE CONVEYOR = ON DO
   BEGIN
     WAIT UNTIL PART_DETECTOR = ON
   PICTURE
   IF XMAX ("HEAT SINK") – XMAX ("CHIP") > = OFFSET TOL
   AND
     ORIENT ("CHIP") WITHIN TILT_TOL OF 90
   THEN
     BAD_PART – OFF
   ELSE
     BAD_PART – ON
 END
```

Assembly application

The Autovision II system has as available output, among others, an RS232 interface, thereby providing simple means for transmitting x, y, θ and part identification information to any sophisticated robot controller which can accept real time information. In the case of the Automatix Cybervision System, the communication path utilised is more direct as the microprocessor of the vision system communicates directly to an identical microprocessor used in the AI 32™ robot controller via shared memory.

Fig. 5. Veryfying hybrid chip location on substrate.

Shown in Fig. 6 is an artist's conception of a complex Cybervision system using multiple cartesian robots, in this case the Automatix AID600™ robot, and solid state cameras (256 x 256) as inputs for the Autovision II Vision System. Fig. 7 shows a simpler application currently used for the mounting of keytops on a customer's keyboard. In this application, the camera is first used to verify that the keys have the correct legend and that there are no lettering defects (100% inspection). If the Autovision System detects a faulty keytop, the robot will pick it

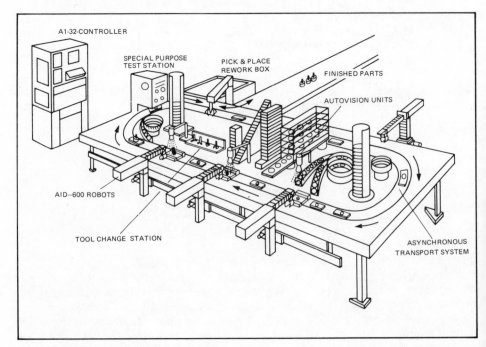

Fig. 6. Multi-robot assembly with Cybervision.

Fig. 7. Cybervision system for keyboard assembly.

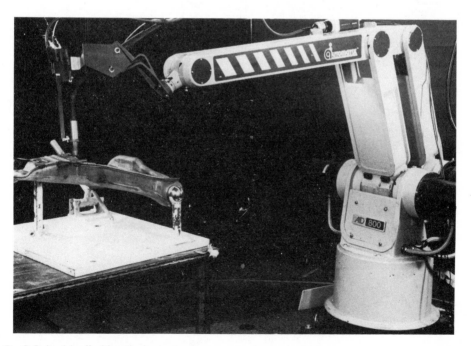

Fig. 8. Robovision II with optical seam tracker.

up and discard it. The vision system can (but presently does not) also verify that the robot has picked up the key and correctly inserted it. If this has not happened, a second cycle will be called for. In this particular application Autovision's capability of recognising the orientation and position of a part is not utilised because of magazine loading.

In another application of the same Cybervision System, the Autovision II is used for the guidance of the AID 600 robot to locate correct positions on a printed circuit board and for inserting non-standard components.

Arc vision welding applications

The Robovision System with seam tracking[5], is shown in Fig. 8. It utilises the same Autovision II subsystem as an integral portion of the Robovision controller. This factory hardened controller is seen on the right hand side of Fig. 9 and contains computer, specialised logic, CRT, keyboard and servo amplifiers. In the robot's arm is the fibre-optic output of the laser light and the fibre-optic input leading to a solid state television camera. In the block diagram of the system as shown in Fig. 10, is the Robovision II seam tracker which, unlike the Kawasaki seam tracker announced last year, is a one-pass rather than a two-pass system.

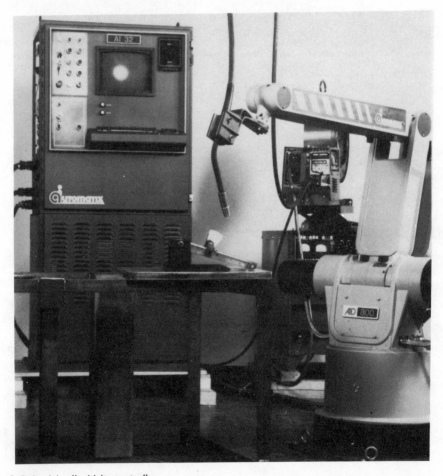

Fig. 9. Robovision II with its controller.

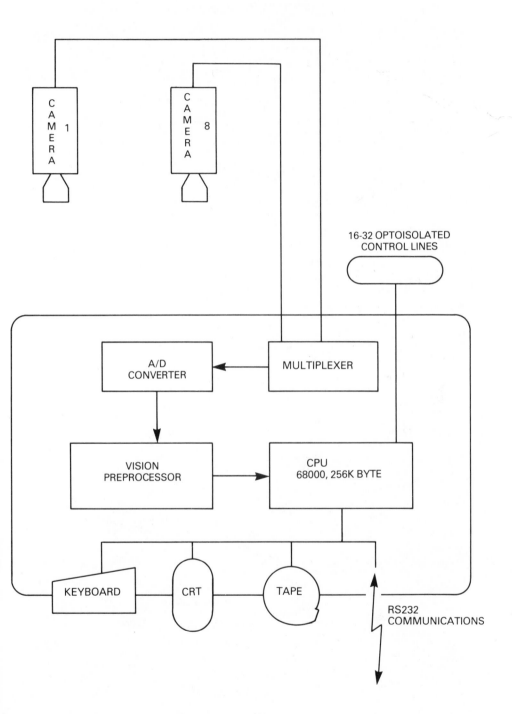

Fig. 10. Block diagram of Robovision II with seam tracking.

Fig. 11. Weldment using Robovision II optical seam tracker.

That is, it observes the welding process while welding occurs, rather than first doing a geometric verification or measurement pass over the workpiece with the vision sensor and then following up with a welding pass. The one-pass system provides not only a speed advantage but ensures that thermal distortions taking place as a result of the welding process are observed and compensated for. The seam tracker scans the area approximately 4 cms ahead of the weld at speeds up to 100 cms/min and measures on the observed image the location of the centre of the seam to be welded, its apparent width, as well as the distance of the workpiece from the sensor.

Deviations of the actual seam beyond preset limits from the nominal or anticipated path result in an error signal to the robot servo system both in elevation and in a direction perpendicular to that of the path. All these calculations are performed with the same Motorola 68000 microprocessor used for other Autovision applications. This results in an updated signal several times per second to the robot. Shown in Fig. 11 is an actual seam weld showing the resulting path with the robot programmed to weld a nominal straight line and the seam tracker providing the error signal to follow the actual desired "S" shaped path. The system is suitable for lap, butt, and fillet welds of a variety of configurations and is not sensitive to weld material or thickness. It does require a visible edge or seam.

It is estimated that at least 30% of all arc welding tasks have not been feasible for robots because of part fit up problems due to thermal distortion of parts, spring back of light metals after forming, or dimensional tolerances such as of flame cut parts. For this reason the use of an optical seam tracker is expected to accelerate the growing acceptance of robot arc welding as a better quality, far less expensive way of doing repetitive production arc welding.

It is believed that as many as 2,000 robot arc welders will be delivered to customers in 1982. Only a handful will have vision based seam tracking, but the percentage should grow rapidly since a seam tracking option averages typically less than one third of the robotic system cost and so is easily economically justified where there is part variability. Seam tracking will also grow because it is more versatile and reliable than other tracking methods developed to date, such as eddy current, contact and through the arc.

Conclusion

Now that programmable vision systems of broad capability are available we will see a rapid proliferation of their use both for high speed 100% part inspection and sorting and for robot guidance, ranging from simple identification and part position determination to true visual servo control where the Vision System is in the servo loop. Most of these applications have been predicted for years in the literature and studied in laboratories throughout the world. Now they have gone on to production applications on the floor of the factory where economic justification, reliability, ease of programming, and robustness are prerequisites.

References

[1] Gleason, Gerald J. and Agin, Gerald J. "A Modular System for Sensor-Controlled Manipulation and Inspection". *Proc. 9th Int. Symp. on Industrial Robots,* Washington, D.C. Society of Manufacturing Engineers, Dearborn, Michigan (March 1979).
[2] VanderBrug, G. J., Ablus, J. S., and Barkmeyer, E. "A Vision System for Real Time Control of Robots", National Bureau of Standards. *Proc. 9th Int. Symp. on Industrial Robots,* Washington, D.C., Society of Manufacturing Engineers, Dearborn, Michigan (March 1979).
[3] Reinhold, A. G., and VanderBrug, G. J. "Robot Vision for Industry: The Autovision System". *Robotics Age, Vol. 2 No. 3, pp. 22-28* (Fall 1980).

[4] Franklin, James W. and VanderBrug, G. J. "Programming Vision and Robotics System with RAIL". Robots VI Conference, Detroit, Michigan, Society of Manufacturing Engineers, Dearborn, Michigan (March 1982).

[5] Libby, C. "An Approach to Vision Controlled Arc Welding". Conference on CAD/CAM Technology in Mechanical Engineering, Massachusetts Institute of Technology, Cambridge, Mass. (March 1982).

A VISUAL SENSOR FOR ARC-WELDING ROBOTS

Takao Bamba, Hisaichi Maruyama, Eiichi Ohno and Yasunori Shiga,
Mitsubishi Electric Corp., Japan

Presented at the 11th International Symposium on Industrial Robots, Tokyo, 7-9 October, 1981.
Reprinted by kind permission of The Japan Industrial Robot Association.

Abstract

This paper describes a new visual sensor to be used for the path correction of arc-welding. By means of an optical pattern projection method using an LED scanning spot, the image patterns of various welding objects have been detected effectively by a two-dimensional photo sensitive position detector with the aid of a microprocessor. The sensing head could be compactly formed so as to be mounted to the robot hand.

The experimental results of this sensor show its sensing abilities to be satisfactory for practical use.

Introduction

In the working environment for program controlled arc-welding robots, the dimensional tolerances of workpieces or fixtures are often beyond their acceptable limits. The sensing capability of path correction is needed in many welding applications. Various sensors to meet such needs have been developed, but their practical use is limited by their inadequate sensing abilities and high cost.

We have developed a new visual sensor system which satisfies the needs of many arc-welding applications. The sensing principle of this system is based on an optical pattern projection method which is usually used for recognition of three-dimensional objects. It can discriminate the welding part of objects by detecting the trajectory shape drawn by a scanning light spot and derive the control information for the path correction.

The sensory head is built compactly. This useful feature is made possible by using a photo sensitive position detector and an LED (Light Emitting Diode) for the main components. These components are also useful for simplification of signal processing which is done by a 8-bit microprocessor (8085A).

Experimental performance of this sensor on some typical objects have shown satisfactory practical results.

Mechanism of weld line sensing

The most familiar direct way of obtaining three-dimensional information is to pass a slit beam of light to an object as mentioned above. Fig. 1 shows the configuration of newly developed sensor and its sensitive area. The main elements of the sensor

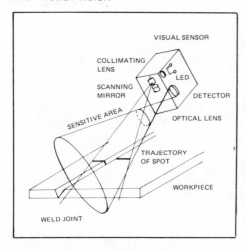

Fig. 1. Configuration of the visual sensor.

Fig. 2. Sensing image in optical pattern projection method.

are a high power infrared LED, a scanning mirror and a two-dimensional photo detector with appropriate optical lens systems. A collimated light beam from the LED is projected onto the surface of workpiece to be welded. Before the beam reaches the surface, it is deflected by the scanning mirror which oscillates at a constant rate. Consequently the beam draws a repetitive linear trajectory on the surface, which provides three-dimensional information of the object when the trajectory is observed from a different direction.

If the weld line is a V-shape groove as illustrated in Fig. 1, the sensing process will be performed as follows (see Fig. 2).

1. As the pattern being picked up forms a trough (point D shown in Fig. 2) at the weld line, this dip in the line trajectory is discriminated by a proper means.

Table 1. Specifications of position sensitive photodetector

Parameters	Typ	Units
Spectral Range at 10% of Peak	300-1150	nm
Peak Response Wavelength	900	nm
Responsivity at Peak λ	0.52	A/W
Position Sensitivity at Peak λ	0.80	A/W/cm
Position Linearity Distortion From Centre to a Point 25% to Edge 75% to Edge	0.3 0.3	% %
Dark Current at 10V Bias	0.5	μA
Rise Time at 10V Bias ($10\% \sim 90\%$, $R_c = 10k\Omega$)	8.5	μsec.
Recommended Mode of Operation	photo-conductive	
Active Area	13×13	mm^2

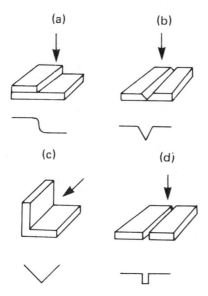

Fig. 3. Various weld joints and sensed patterns.

2. The horizontal deviation of the desired welding point (D in Fig. 2) from the centre of sensing plane is calculated.
3. The deviation signal is fed to a robot control system for path correction.

The visual sensor also has a distance sensing capability between the sensor and the workpiece. As the observing axis slants against the light projecting axis, the basic line of trajectory $(\overline{AC}, \overline{EF}$ in Fig. 2) moves up and down according to the relative distance. In particular the Y value of the spot position B (Y_0 shown in Fig. 2) represents the relative distance along the observing axis. The distance (l) can be calculated by the following formula:

$$l = l_0 (1 - \alpha Y_0)^{-1} \qquad (1)$$

where l_0 = the distance between the centre of the lens and the point where the light projection axis crosses the observing axis.
α = a constant value defined by the device arrangement and lens systems.

The principle of optical pattern projection onto a workpiece has the further advantage of its sensing ability. Fig. 3 shows various types of weld line used for arc-welding and their sensed images. Since each pattern has a break point on its joint part, every weld line can be easily detected by this sensor.

Fig. 4 shows a prototype of our visual sensor. It can be of compact size by employing a LED as a light source, which makes the sensor easily attachable to the robot wrist.

Detecting device

In general, image sensors such as ITV or CCD cameras are usually employed as two-dimensional optical sensors in robot eyes. However they have some problems as follows:
1. Appropriate illumination control must be needed according to the detecting situations.

Fig. 4. A prototype of the visual sensor.

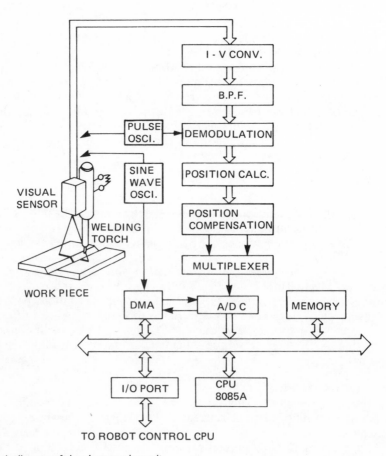

Fig. 5. Schematic diagram of signal processing unit.

A (SENSED PATTERN)

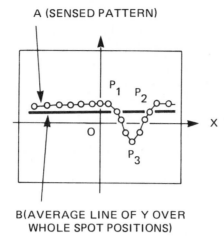

B(AVERAGE LINE OF Y OVER
WHOLE SPOT POSITIONS)

Fig. 6. Method of weld line detection.

2. It takes a considerable length of time to process the image information from a large number of pixels for real time use.
3. Total system cost might be expensive for practical use because high performance processing equipment is required for pattern recognition.

These features restrict the practical applications of image sensors.

To avoid these problems, we recently adopted the two-dimensional lateral photoeffect diode as the area sensing device. It is a dual-axis position sensitive detector that provides continuous X- and Y-axis positional information of a light spot movement on the detector. Since the device is not divided into pixels, no internal scanning is required. Thus the position of the spot can be detected immediately. Also, the peripheral circuits and image processing algorithms for the sensing system can be simplified. Table 1 shows the main specifications of this device.

Signal processing circuit for the sensor

Fig. 5 shows the schematic diagram of signal processing circuit for our sensing system. The LED is modulated by a pulse oscillating unit ($f = 10$ kHz). The detector picks up the reflective light from workpieces, and outputs continuous current signals which carry the position information of the moving spot image on the detector surface. Through current-to-voltage converters, the signals are fed to band-pass filters ($f_c = 10$ kHz) to extract effective LED signals from the background light noise. After demodulation, accurate X and Y position signals of the detecting spot are separately derived by operational amplifier circuits compensating the influences of its intensity.

An 8-bit microprocessor (8085A) is used for weld line discrimination from the sensed waveform. After the compensation described above, two-channel signals (X, Y) are digitised by an analogue-to-digital converter. A direct memory access controller (DMA controller) transfers this digital position information to random access memories (RAM) of the CPU synchronously with the mirror scanning cycle. Thus one frame of spot trajectory information is stored in the memories rapidly. Refresh cycle of the spot trajectory is 10 ms.

The microprocessor perceives the deviation of the weld line and the relative distance between the workpiece and the visual sensor. In the following section, algorithms on the perception will be described.

Configuration of software

Algorithms for weld line perception are simple. Assuming that the observing axis of the visual sensor is located perpendicularly to the surface plane of the workpiece, the waveform can be processed easily. If the weld line is a V-shape groove, the processing procedures are as follows (see Fig. 6):

1. The average value of the Y coordinates over a whole spot position is calculated, which gives line B in Fig. 6.
2. Crossing points of the waveform (A) to the average line (B) are obtained. These points are P_1 and P_2 in the figure.
3. The point of maximum deviation from the average line is obtained (point P_3).
4. The desired point is regarded as that of maximum deviation between P_1 and P_2. In this case this point is P_3.

By using the average value, the required weld point can be detected correctly even if the waveform is disturbed by noise. Moreover this method is applicable to

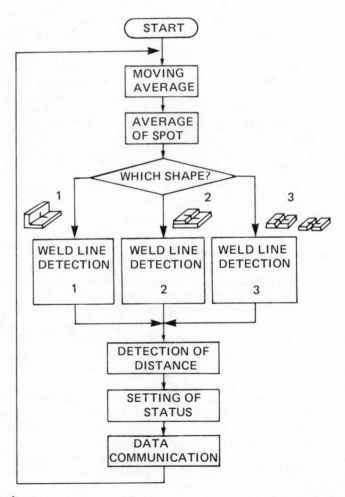

Fig. 7. Flowchart of software.

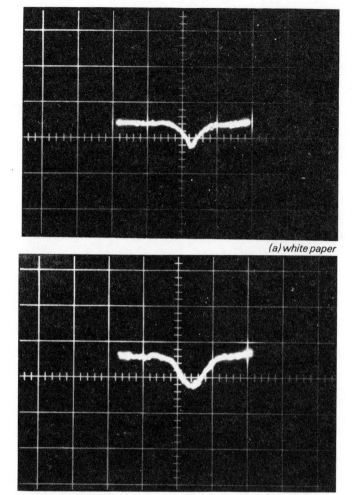

(a) white paper

Fig. 8. Sensed patterns.

(b) rusted iron

other types of weld line. For example, in the case of a weld line such as (a) in Fig. 3, the weld point is regarded as the crossing point of the pattern to average line. In the case of (d) in Fig. 3, it is regarded as the midpoint of the crossing points.

The configuration of software is shown in Fig. 7. A first noise reduction using the method of moving averages is done. That is:

$$W_i = W'_i + (B'_i - W'_i)/N \quad (i = 1, \ldots, m) \tag{2}$$

where W_i = i-th spot position in the present frame of the image.
W'_i = i-th spot position in the previous frame of the image.
B_i = i-th spot position of newly detected data.
N = sampling number.

This process can remove random noises from the sensed image. Secondly weld line discrimination is done. A proper discriminating routine is selected corresponding to the shape of weld line.

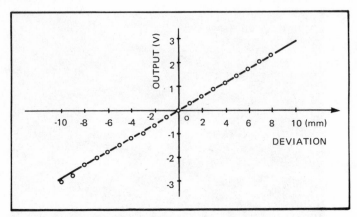

Fig. 9. Sensor output in position sensing of weld line.

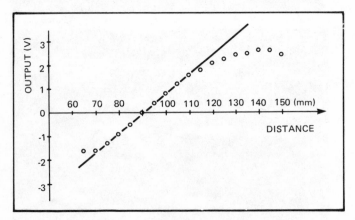

Fig. 10. Output in distance sensing.

Next, the distance between the visual sensor and the workpiece is obtained by the calculation shown in section 2.

In the status setting stage, the microprocessor checks the sensing status. For example it checks whether the light beam is correctly projected onto the workpiece, whether the given information is normal, and whether a weld line is observed in the field of view. This status can be used for safety control.

The processing time for a one cycle execution of the software is less than 40ms.

Experimental results

The sensing ability of optical sensors depends on the optical characteristics of the workpiece surface. Unfortunately the materials used in arc-welding usually have unfavourable reflective characteristics. This makes it difficult to detect the correct spot position on the workpiece. In addition, the sensing distance will be limited. We have done experiments on such problems with our visual sensor.

Fig. 8 shows the experimental results of image detection for a V-shape groove using white paper (a) and rusted iron (b) respectively. Our visual sensor detects a correct image even in the worse condition such as the heavy rusted iron. In further experiments, we have ensured its sensing capability to other materials, i.e. stainless steel, aluminium and so on.

Sensing linearity to a typical workpiece is shown in Fig. 9 in which the abscissae indicates the deviation of weld line from the centre of the visual sensor, and ordinate is the sensor output. As the field of view is approximately 20 mm × 20 mm at 90 mm distance, it can be said that the sensed weld line deviation is linearly proportional in almost all of the area.

The next experimental result is the sensing capability of the distance between workpiece and sensor. Fig. 10 shows the typical response for rusted iron. The output of the visual sensor has a good linearity in the range from 70 mm to 110 mm, but it deviates from an ideal curve in other domains – nearer than 70 mm and further than 110 mm. In the former it is becoming outside the field-of-view and in the latter there is a shortage of input power.

Although these limitations for sensing abilities exist, the visual sensor is effective for almost all of workpiece materials and weld line shapes with a sensing resolution of less than ±0.5 mm.

Conclusions

The visual sensor using the optical pattern projection method has been reported. It has the following distinctive features:

1. Three-dimensional position sensing ability of weld line.
2. Simple sensing mechanism.
3. Sensing ability to various weld line shapes and materials.
4. Fast signal processing facilities compared with other systems using an image sensor such as an ITV.
5. Compact size easily attachable to a robot hand.
6. An adequate sensing resolution of less than ±0.5 mm.

We have confirmed experimentally its effective sensing performance which will be of great use for practical intelligent robots.

VISION GUIDED ROBOT SYSTEM FOR ARC WELDING

Ichiro Masaki, Unimation–Kawasaki, Joint Development Team, USA,
Maurice J. Dunne, Unimation Inc., USA and
Hideo Toda, Kawasaki Heavy Industries Ltd., Japan

Abstract

The vision guided system for arc welding has the function of sensing and following the seam to be welded. The robot repeat mode for every successive workpiece consists of a sensing path mode and a welding path mode. In the sensing path mode, the robot arm picks up the sensing hand and carries it over the seam to sense the position of the seam to be welded. Welding is performed in the welding path mode, with the welding hand, based on the positional information which was gathered in the sensing path mode.

Introduction

In many actual production lines, the positional accuracy of the seam to be welded is not good enough to use robots which do not have seam following functions. The robot which can follow the seam does not require expensive quality control, to keep the position of the seam accurate enough, of the parts to be welded. The seam following function is also useful to increase the quality and the reliability of automatic arc welding.

An effort to transfer a vision guided robot system for arc welding from the laboratory to the real world has been carried out. The system was improved for higher reliability, easier maintenance, and better cost performance.

System operation

The system operation consists of the following procedures as shown in Fig. 1:
1. Teaching mode.
2. Training path mode.
3. Sensing path mode.
4. Welding path mode.

In the teaching procedure, the program and the location data are taught. The program is written with the VAL Language which was developed for robots. The VAL Language includes instructions for usual arm motions, communications with peripherals, program editing, mathematical calculations, and welding. The program is input through a CRT/keyboard terminal and stored in the memory of the VAL controller. A mini floppy disk is available as the second storage device. The location data are usually taught by using a teach pendant. For example, the

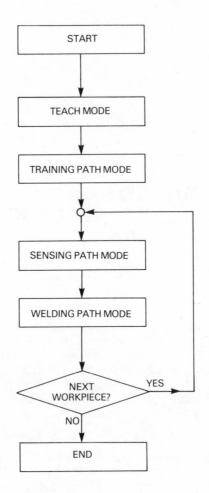

Fig. 1. System operation.

location data for point A can be taught by moving the robot arm to the position of point A with the teach pendant and keying in HERE A. The teach pendant has switches not only for every joint of the arm, but also for every axis of the world rectangular coordinates and the tool-oriented rectangular coordinates. This teaching procedure is carried out only once for each welding program.

In the training path mode, the robot arm picks up the sensing hand and carries it over the seam. The training path is performed in order to store the template, or reference, images and the template image table. Because the template image depends on the shape of welding joint, at least two different template images should be stored, for example, when it is required to perform both butt welding and corner welding in the same robot repeat cycle. The template image table tells what template image should be used with what seam in the present particular application. In the training path, the image is acquired at every taught point. When the difference in shape between the acquired image and the last template image is larger than the preset threshold level, the acquired image is stored as a new template. A mini floppy disk is available as the second storage to store the template images and the template image table. Like the teaching process, this training procedure is carried out only once for each welding program.

After performing the training path mode automatically, the system is ready for the robot repeat mode. The robot repeat mode, unlike the teaching and training mode, is repeated for every successive workpiece. It consists of the sensing path mode and the welding path mode.

The robot arm picks up the sensing hand and carries it over the seam to obtain images, in the sensing path. The points where the sensing should be performed are decided in the program. The instruction to sense every 10 mm is SENSE 10. This instruction can be used to decrease the total number of taught points and to make the teaching easy. The image processor compares the acquired image with the template image and outputs the positional deviation. The proper template image for this processing is chosen based on the template image table. The output of the image processor is stored in the VAL controller.

After performing the sensing, the sensing hand is replaced with the welding hand automatically to perform the welding. The hand/hands which is/are not attached to the robot arm is/are kept by the hand holder. The hand holder has nests for the sensing and welding hands and clamps the hand(s) pneumatically based on signals from the VAL controller.

In the welding path mode, the positional destinations for welding are calculated, by the VAL controller, based on the information gathered during the sensing path.

System components configuration and function

The block diagram of the Vision Guided Robot System is shown in Fig. 2. The sensing hand is to gather images and to transfer them to the opto-electronic transducer. The main components of the sensing hand are an optical slit pattern projector, an objective lens, and a fibre optic cable. An optical slit pattern is projected onto a seam and is deformed according to the shape of the welding joint as

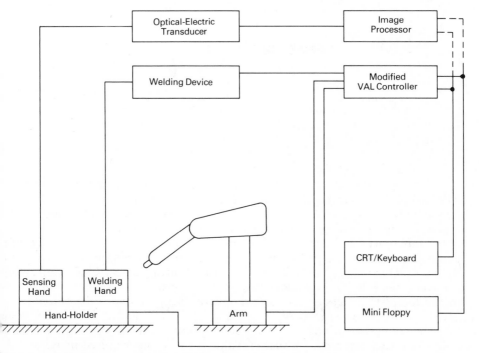

Fig. 2. System architecture, vision guided robot system.

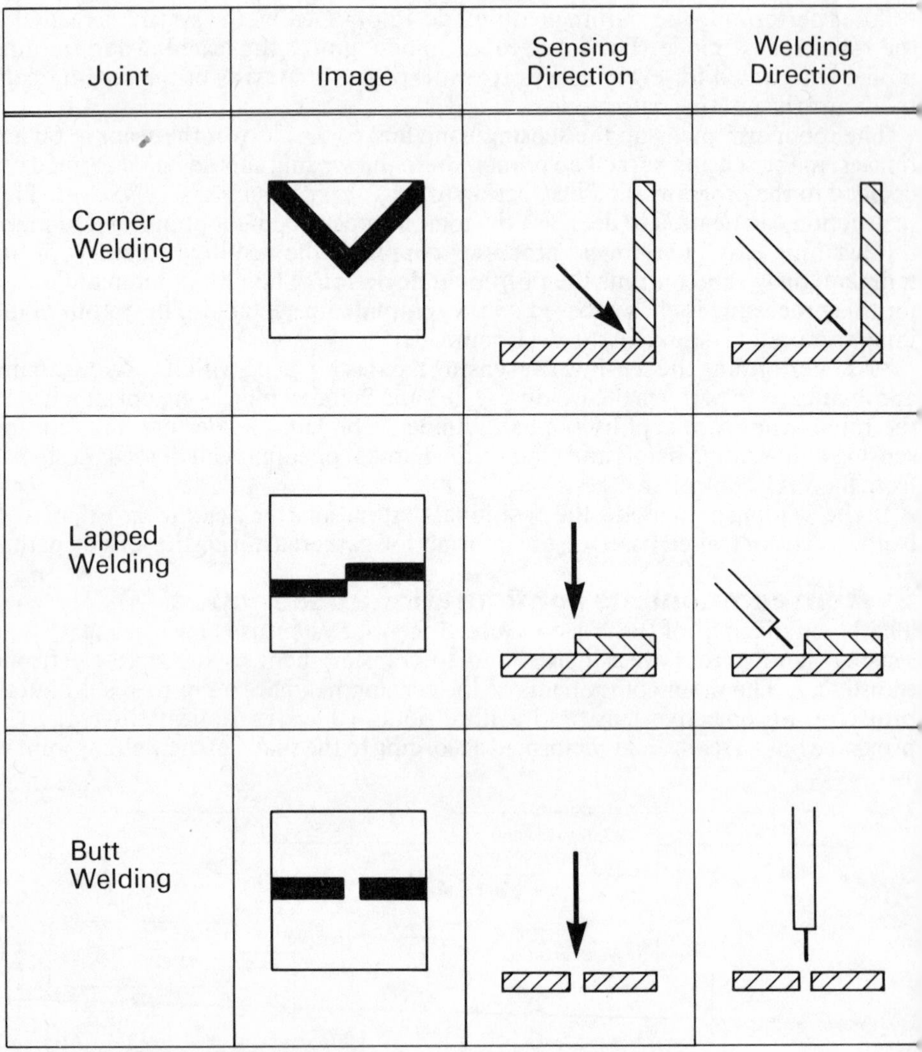

Joint	Image	Sensing Direction	Welding Direction
Corner Welding			
Lapped Welding			
Butt Welding			

Fig. 3. Joint geometry.

shown in Fig. 3. The deformed optical pattern images are gathered by the objective lens and are transferred to the opto-electronic transducer which is mounted remotely. A solid-state TV camera is used as an opto-electronic transducer and it transduces the optical signal into the electric signal for the image processor.

The function of the image processor, in the training path mode, is to store proper templates. In the sensing path mode, the image processor outputs the positional deviation. Our image processing method was named "Direct Cross-Correlation Method", because the cross-correlation values between an acquired image and shifted template images are calculated directly. Suppose that the acquired image and the template image are in the X-Y plane. If the shifting value $(\triangle X, \triangle Y)$ is equal to the positional deviation between the acquired image and the non-shifted original template image, the cross-correlation value between the acquired image and the particular shifted template image which is shifted by $(\triangle X, \triangle Y)$ is the highest. The

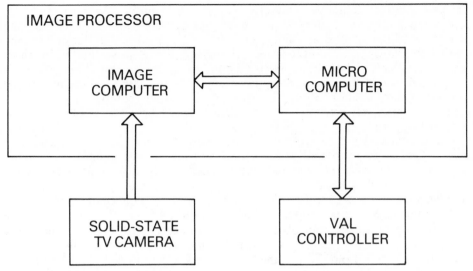

Fig. 4. Image processor.

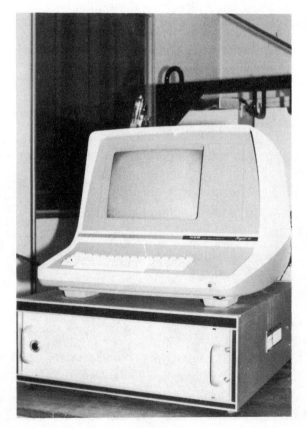

Fig. 5. Image processor.

image processor calculates many cross-correlation values between the acquired image and many shifted template images which are created from the same template with many different shifting values, and detects the positional deviation between the acquired image and the non-shifted original template image by finding the shifting value which makes the cross-correlation value the highest. This process is performed by our image computer, shown in Fig. 4, in order to realise high speed processing. The image computer is supervised by a micro computer. The sensing hand, the optical-electric transducer and the image porcessor work in the training and sensing path.

On the other hand, the welding hand and the welding device work in the welding path. The welding hand has a welding torch which is connected electrically to the welding power supply. The robot arm has a socket which can grip either the sensing hand or the welding hand. The hand-holder holds the hand(s) which is/are not attached to the robot arm. Both the arm socket and the hand-holder are actuated pneumatically and are controlled by the VAL controller.

The VAL controller controls and/or communicates with the arm, the image processor, and the peripherals. In the teaching mode, the VAL controller stores the program, which is input through the CRT/keyboard terminal, and moves the arm based on the manual input via the teach pendant. In the training mode, the VAL controller tells when the image should be acquired. In the sensing path mode, the VAL controller tells the image processor when the image should be acquired, and stores the output of the image processor. The positional destinations for welding are calculated by the VAL controller during the welding path mode.

The CRT–keyboard terminal is used by the operator to communicate with the VAL controller or the image processor. The mini floppy disc drive stores programs and template images into floppy disks and also loads into the VAL controller or the image processor. The welding torch and the welding power source can be selected for every application.

Features
Several features of this system are listed below.

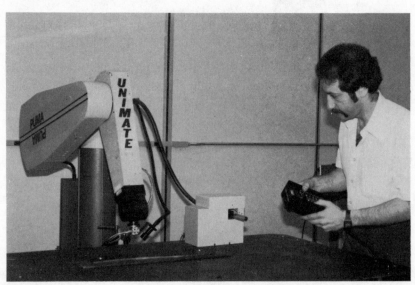

Fig. 6. Robot arm.

Direct Cross-Correlation Image Processing Method

Our direct cross-correlation image processing method has the merit of being little affected by optical noise and the disadvantage that it requires a long processing time with the usual Von-Neumann computer. This disadvantage was eliminated by developing an image computer as mentioned below. High reliability is realised through relative immunity to optical noise.

Image Computer

High speed image processing is important for quick sensing. Our image computer was developed for inexpensive and high speed image processing.

Tack Weld Override

In many cases, the seam has tack welded parts. Our system can recognise tack welded parts and can override them.

Automatic Template Image Acquisition

It is not required to store template images manually. In the training path mode, the system acquires images and makes the decision for every acquired image as to whether it should be stored as a template image or not.

VAL Language

The VAL language has been used as a convenient language for robots. It was improved to communicate with the image processor and to calculate the positional destination data for welding.

Specifications

The specifications for the first production version are listed below. Other versions are available for the user's convenience. For example, the PUMA 750 robot arm will be used, instead of the PUMA 600 robot arm, for larger working space and heavier load capacity. The following specifications are also subjects for continuing improvements.

Sensing range: ± 10mm, in horizontal and vertical directions
Sensing resolution: 0.4mm
Sensing speed: 17 points/second (in the typical condition)
Welding speed: Depends on the specifications of the welding device, the shape of the seam, etc. The maximum speed of the PUMA 600 robot arm is 500mm/sec.

Acknowledgement

This development has been continued in the joint development team of Unimation Inc. and Kawasaki Heavy Industries Ltd. Both companies have had a strong tie-up for thirteen years. The joint team is in a Unimation research and development facility which is located on the east coast of the USA. Colleagues in both companies have supported this study strongly.

Previous publications

Masaki, I., Toda, H. 'Kawasaki Vision System', Procs. 10th International Symposium on Industrial Robots, Milan, IFS (Publications) (March 1980).
Masaki, I., Dunne, M. J., Toda, H., et al. 'Welding Robot with Visual Seam Tracking', Procs. Developments in Mechanical Automated and Robotic Welding, London, The Welding Institute (November 1980).
Masaki, I., Dunne, M. J., Toda, H., et al. 'Arc Welding Robot with Vision', Procs. 11th International Symposium on Industrial Robots, Tokyo, Japan Industrial Robot Association (October 1981).

PROGESS IN VISUAL FEEDBACK FOR ROBOT ARC-WELDING OF THIN SHEET STEEL

W F Clocksin*, P G Davey*, C G Morgan, and A R Vidler,
University of Oxford, Oxford, UK

Presented at the 2nd International Conference on Robot Vision and Sensory Controls – Stuttgart,
2-4 November, 1982

Abstract

Using conventional robots for MIG (Metal/Inert-Gas) arc welding on thin (1 to 2 mm) sheet steel pressings is restricted by the difficulty of maintaining accurate fit-up and fixturing. Dimensional variations are introduced by wear of tools and fixtures, effects of variable springback in sheet steel pressings, and thermal distortion during welding. These variations cause errors in the position of the welding torch with respect to the seam of up to ± 3.0 mm, which prevents the formation of good welds. This paper gives a very brief description of our first implemented system that forms and uses semantic models for the automatic visually-guided control of a MIG welding robot. The system uses its models to detect the following dimensional variations in thin sheet steel assemblies: gap width, stand-off error, and lateral error. Lap-, T-, and butt-joints are dealt with. Due to the semantically-based visual feedback technique reported in this paper, errors in the position of the welding torch are corrected to less than ± 0.3 mm, which is a suitable tolerance for producing good welds in such pressings. Furthermore, gap widths are detected for the purpose of dynamically altering welding parameters.

Introduction

In a previous report[1] we reviewed the particular problems of using conventional playback robots for MIG welding of 1 to 2 mm thin sheet steel pressings, and the reasons for them. We analysed the relative costs likely to be incurred by a motor car manufacturer when solving these problems on a typical assembly using two possible methods: first, by using conventional robots, but reducing dimensional variation by conventional production engineering techniques; and second, by fitting the robots with sensors to correct both the programmed path and welding parameters. It emerged that the capital and recurrent costs of using robots with

GEC Fellows in Robotics at St Cross College, University of Oxford. The Robot Welding Project is supported jointly by the Science and Engineering Research Council, and BL Technology Ltd.

sensors would be about a fifth of that using more precise tooling; to develop appropriate sensors is clearly justified by cost alone.

We now report an implementation of semantically-based visual feedback for visual guidance of a MIG welding robot. We use the term 'semantic' because the system forms and uses models of visual and geometric relationships of the workpiece. Differences between template workpiece models and production workpiece models are 'explained' in terms of specific fit-up and jigging variations, and give rise to error signals that are fed back to control the robot. Error signals control not only positional parameters but also welding parameters.

The implementation reported here is preliminary. Trials thus far have been made on joints between planar surfaces, and with inspection and welding being done in separate passes of the seam. Although the sensor functions within close proximity to the arc, the image processing operations are at present too slow to permit simultaneous inspection and welding.

System overview

The development system consists of a conventional robot equipped for MIG welding, a control subsystem, and a vision subsystem. At present, the robot is an ASEA IRb-6 industrial robot with five degrees of freedom and a payload of 6 Kg at rest. The IRb-6 is equipped with an Aga MIG-welding set, and both robot and set are controlled by the standard ASEA controller containing an I8008 microprocessor. We have made minor interface modifications to the welding set and 8008 program.

The control sub-system is only briefly described in this report. It consists of an LSI-11/23 which is connected to the ASEA controller by a 9600 baud serial line. As the ASEA controller is not equipped with a coordinate conversion capability, we have implemented a kinematic model in the control sub-system to permit

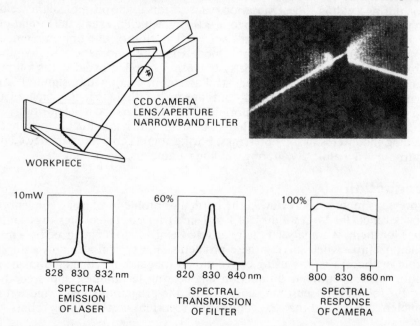

CCD CAMERA
LENS/APERTURE
NARROWBAND FILTER

WORKPIECE

10mW

828 830 832 nm

SPECTRAL
EMISSION
OF LASER

60%

820 830 840 nm

SPECTRAL
TRANSMISSION
OF FILTER

100%

800 830 860 nm

SPECTRAL
RESPONSE
OF CAMERA

Fig. 1. The laser light sheet is projected across the workpiece seam to form an image similar to that shown on the right. As shown by the graphs, the interference filter is intended to reject wavelengths other than those emitted by the laser.

programming and executing ASEA programs in a number of different cartesian "world" coordinate systems. Our implementation of coordinate conversion works in real-time, does not exhibit cumulative errors over successive forward and inverse conversions, deals with arbitrary offsets of multiple tools about the wrist, is easily calibrated by the user, and permits arbitrary geometrical transformations and following of arbitrary space curve paths, interpolating in world coordinates. Because the sensor subsystem is considered as a tool, the control subsystem also deals with arbitrary transformations between image coordinates, and corresponding robot joint coordinates.

The sensory sub-system consists of another PDP-11/23 computer equipped with a sensor and interface. The sensor, consisting of a rectangular CCD array camera, a GaAlAs infra-red laser diode emitting at 830 nm, and a narrowband optical filter, are arranged as shown in Fig. 1. The camera generates a PAL standard video signal which is digitised and stored as a 256x256 array of 8-bit pixels. A line ten pixels in length subtends approximately 1 mm on the focussed inspected surface. The sensor assembly is fixed to the end-effector of the robot, near the welding torch. Output of the CCD chip appears little affected by spurious pickup of the electromagnetic radiation from the arc, and fitting the camera with a narrowband optical filter having a spectral halfwidth of 10 nm centred over 830 nm effectively removes much of the visible arc emission from the image.

The sensor operates on the principle of active triangulation ranging, a technique which many researchers have used for obtaining information about the three dimensional layout of surfaces in a scene, for example, Agin and Binford[2]; Popplestone and Ambler[3]; Hill and Park[4]. The laser diode emits through a narrow slit and cylindrical lens to cast a sheet of light on the workpiece. The intersection of the light sheet and the workpiece surface, when viewed from a different regard with the camera, forms a curved or straight-line stripe that depicts the layout of a particular cross-section of the surface relief of the workpiece.

The images shown in Fig. 2 result when the light sheet crosses different types of weld joints. The nearer to the top of the image that a point on the stripe lies, the further away is the surface to which the light is projected. The purpose of the sensor described here is not to examine the molten weld pool, but to provide a means of determining the geometrical relationship between the torch and the seam. The actual image processing operations used are discussed in the next section.

Currently we use this apparatus to enhance the conventional robot teaching/welding sequence by performing the following three phases:
1. Manual teaching of weld seams on a pre-production master assembly. Straight-line seams are more easily taught by specifying a pair of endpoints. The kinematics subsystem automatically derives the straight line in space connecting the endpoints*. Furthermore, weaves can be taught by specifying a weave amplitude: the kinematic model then automatically derives a sinusoidal weaving path directed along the seam, and lying in the plane normal to the torch.
2. Automatic visual survey of taught points on the master assembly. The control system runs the taught robot program unattended on the pre-production master assembly, and uses the sensor subsystem to identify several items of reference information for each taught point, including: the type of weld joint at each

---

* It is a simple matter to extend the system to interpolate circular paths from three points and curved spline paths from four or more points. More importantly, perfectly taught trajectories will not be necessary when the sensory subsystem is able to adjust the trajectory to fit the workpiece at hand.

point, the position of the torch relative to 'terrain' near the seam, and a metal thickness factor for lap seams. Gap width is known to be negligible on master assemblies. The system also generates its own idea of where the torch ought to be positioned, but this capability is not used in the current implementation. Currently, this entire step takes about 2 seconds per taught point. For a single straight-line seam, this amounts to at most 5 seconds to survey the seam. It is likely that a later implementation will merge this inspection phase with the previous manual teaching phase.

3. Welding with visual correction of taught points and welding parameters. This phase is repeated for each production assembly, and it currently involves making two passes over the workpiece: inspection (3a) and welding (3b). For phase 3a, the control subsystem drives the robot to the taught seam endpoints. At each endpoint, the sensor subsystem measures the width of the seam gap, and the position of the torch relative to the seam. This information is compared with the reference information derived previously from the pre-production master assembly. Discrepancies between the master and the production assembly are treated as error signals. The error signals are fed back to the control subsystem, which automatically derives a new corrected program to weld the seam (3b). To save time, the inspection pass follows the taught program along the seam, and the welding pass follows the corrected program in the reverse direction.

The following error signals are detected and corrected: stand-off error (torch altitude above the seam along the torch axis), lateral error (deviation across seam from centre of seam), and gap width. The gap width and (known) metal thickness are used to calculate the wire feed speed and torch travel speed using a parametric welding model[5] as modified and extended by Vidler[6]. At the moment the time taken for each seam is about 2 seconds per point analysed plus the time taken for the actual welding.

Several points about the above procedure bear mention. First, metal thickness must be input to the system by the operator. This is a reasonable condition, as

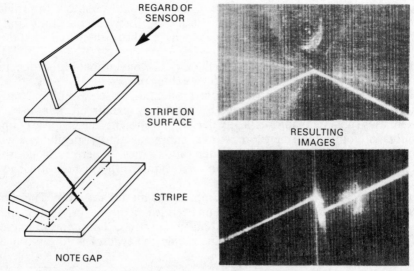

REGARD OF SENSOR

STRIPE ON SURFACE

RESULTING IMAGES

STRIPE

NOTE GAP

Fig. 2. A T-joint and lap-joint are shown here together with the characteristic stripe pattern received by the sensor. These images are processed by software to produce segment descriptions which are then used to identify the joint and its attributes.

metal thickness is specified when the assembly is designed, and it is known to the operator. Visual detection of a metal thickness factor during Phase 2 above is used for the purpose of detecting gaps in lap joints, described below.

Second, path correction and welding, Phase 3 above, forces the system at present to make two separate passes over the seam: an inspection pass (3a) and a welding pass (3b). The consequence is that variations due to thermal distortion during welding are ignored. This has proved not to be a problem in our trials, but simultaneous inspection and welding would be required to increase the yield of correctly welded seams in a production environment. The reason our initial implementation performs correction and welding in separate passes is that the current image processing algorithms are implemented wholly in software, and at 2 seconds to analyse an image, this is too slow for adequate simultaneous closed-loop control. For a maximum travel speed of 20 mm/s, and correcting every 2 mm along the seam, an image processing time of less than 100 ms is required. We are currently developing a more specialised image processor that should enable the system to attain this cycle time. Finally, although the image processing algorithms have been tested in the presence of arc emissions with little ill effect, more exhaustive tests need to be undertaken to determine whether the algorithms need to be more robust.

Vision representations and algorithms

The software for the sensor subsystem is written mostly in Pascal[7]. The Pascal compiler we use generates PDP-11 machine instructions, but about 10 per cent of the system is hand coded in PDP-11 instructions to decrease the time consumed for image processing. Three levels of representation are used by the sensor algorithms. The lowest level is the image itself, which is represented as a 256×256 array I of integers in the range 0 to 255 called pixels. Pixels are addressed by a row index and a column index, and the pixel at row r and column c is denoted $I(r,c)$. The next level of representation is the "hit list", which is a 1×256 array H of row indices in the range 0 to 255. The ith element of H is denoted $H(i)$. Given an image I containing a stripe as described in the previous section, $H(i)$ is defined to be the row index where the stripe in the image intersects the ith column of I. If no part of the stripe intersects column i, then $H(i)$ is undefined (assigned a unique value). Array H can be considered as a partial function of i that represents the stripe, as the stripe can intersect a given column no more than once.

The final level of representation is an array S of segments, where each segment represents a straight-line segment lying in I. Components of a segment are the row and column indices of its endpoints, its length, and its slope. A stripe in the image is represented as an array of segments.

If two points lie on the image of the stripe, then the distance between the points on the workpiece surface to which these image points project can be calculated. Scaling factors (Sr, Sc, Sz) having units of mm/pixelwidth are automatically calculated when the system is calibrated, and they are assumed constant everywhere within a certain working area of the image.

The sensor software provides functions for capturing images of stripes, transforming one representation into another, storing and comparing segments, and computing the error signals that are fed back to the control computer. Consider first the image processing operations carried out for teaching seam endpoints (Phase 1 of the previous section). First, the robot drives the camera so that the endpoint of the seam is in the centre of the image. This is done by commanding the control subsystem to drive the camera so that the optical axis is co-linear with the axis of where the welding torch would be if the tip of the torch was positioned at

the endpoint, normal to the seam. Provided that the fixed relationship between the welding torch axis and the optical axis is known, then the kinematic model is responsible for carrying out this transformation and robot movement. Next, an image is captured into array I, and the sensor subsystem must now carry out some computations and store in a database some information associated with the given endpoint. The computations are divided into five steps summarised as follows:
1. Construct the "hit list". This reduces the amount of information that encodes the stripe from the 65536 integers in the image to the 256 integers in the hit list.

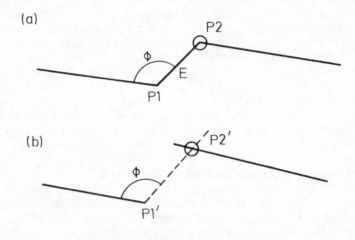

Fig. 3. (a) Segments obtained from a pre-production master assembly used as a model lap-joint. The circle denotes the calculated target location for the welding torch. (b) Segments from a lap-joint having a small gap. The dotted line is hypothesised from the model, and the target location is recalculated. (c) The four possible characteristic patterns of segments resulting from lap joints.

2. 'Clean' the hit list to remove noise and hypothesise missing data.
3. Fit line segments to the hit list. This reduces the information still further to about one to ten segments.
4. Identify the type of joint from the configuration of line segments. This locates the seam and represents the joint as a single label. The joint is associated with the endpoint.
5. Determine the vector offset from the torch to the seam. This computation is actually performed by the control subsystem, since it involves transformations from one coordinate system to another.

As we are considering Phase 1, the vector offset is a relationship determined during teaching, so it will be associated with the endpoint as the 'ideal' offset to be repeated for the production assemblies.

For inspecting production assemblies prior to welding (Phase 3a of Section 2), the same steps as above are followed. However, once the vector offset is computed, it is compared with the 'ideal' offset in the database previously computed when the seam was taught on the pre-production master assembly (Phase 1). Any discrepancy is computed and the difference is fed back to the control sub-system as a positional error. The error is used by the control sub-system to construct the corrected welding program. We now discuss the image processing steps in detail.

Constructing the Hit List

To obtain the element H(i), the pixels in column i of the image are convolved with a discrete linear filter mask. The row index where the maximum convolution response occurs over column i is assigned to H(i). This process is repeated for each column in the image.

Several criteria govern the choice of filter mask. The primary criterion is that the filter should give a maximum response to the cross-section of a stripe, thus discriminating the stripe from other sources of brightness such as noise, mutual illumination of one steel sheet by reflection of laser light from another, and any molten metal splatter shooting through the field of view having spectral components in the range 825-835 nm. In examining typical images, we found that brightnesses of non-stripe regions in the image could be as bright as the stripe. Furthermore, the absolute brightness along the stripe could fluctuate across the entire brightness range of the image. Thus we ruled out simple thresholding and the use of histograms to select thresholds. We found that it was reasonable to assume that the width of the cross-section of the stripe (sampled along an image column) was defined to within fairly narrow bounds, and that the width did not change significantly when the slope of the stripe was non-zero. Thus we were led to using differencing techniques, and could assume a fixed maskwidth. Four techniques were implemented and tested, of which one survives in the present system. Each column of the image is convolved with a discrete approximation to the DOG (difference of gaussian) filter[8]. This particular discrete approximation, which is equivalent to a second differences filter of zero net gain, consists of a central positive region (width 3, gain 2) surrounded by a negative region (two each: width 3, gain −1). The DOG and its family of discrete approximations is sensitive to impulses having widths less than the width of the mask's central region. Although the spatial frequency response of the DOG filter is well localised, large amplitude impulses have significant power not only in the high spatial frequencies but also in the portion of the spatial frequency spectrum where the DOG filter is normally tuned. In spite of this limitation, which has not yet caused problems, the current implementation of the sensor sub-system uses this filter. Given this mask, convolution requires no multiplications, only nine additions and one binary shift.

The system also improves performance by convolving only a certain sub-range of each image column, using a statistic based on previous hit values to define the centre and length of the subrange for a given image column.

Cleaning the Hits

It is possible that hits can be incorrectly assigned by the hit detection filter, for example, when a non-stripe pulse in a given column appears more "stripe-like" than the stripe pulse itself. The causes of such noise were mentioned above, and it is necessary to process the hit array to limit the effects of such noise. First the hit array is cleansed by removing noise points and then connecting small gaps if they seem to lie on the stripe. Distinguishing between hits correctly positioned on the stripe and those misplaced is made easier because the hit array encodes depth information, and tolerances on possible surface profiles are known. For example, any $H(i)$ that differs with one of its neighbours by more than a threshold is considered "noise", because the smoothness (continuity) of the metal surface is the limiting factor. Any such noise elements are replaced by an "undefined" marker. Furthermore, any connected run of hits less than 10 pixels (1mm on the metal surface) is marked as undefined, as it is impossible for any physical feature on the metal surface to give rise to such a pattern of hits. The result of these two operations is to leave a pattern of connected hits and connected runs of undefined elements (gaps).

The next step examines each gap to account for the possibility of drop-outs along the stripe. The pixel values of the two defined endpoints of the gap are

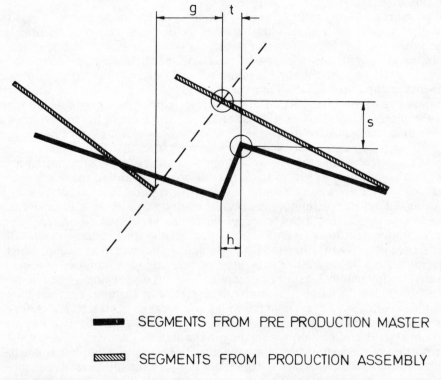

■■■ SEGMENTS FROM PRE PRODUCTION MASTER

▨▨▨ SEGMENTS FROM PRODUCTION ASSEMBLY

Fig. 4. Two lap-joint segments are shown overlaid to depict the derivation of gap width g, standoff error s, lateral error t, and metal thickness factor h.

examined, and a threshold is defined as 70% of the minimum brightness of the two endpoints. Then, each pixel on the straight line between the endpoints is compared with the threshold, and is reclassified as a hit if the pixel's brightness exceeds the threshold. This operation serves to hypothesise hits within gaps in the cleaned hit list, and it is the last operation in the processing sequence to examine the image array. In practice we have found that the number of pixels examined in this way varies from none to about a dozen.

Fitting Segments
Next, the system fits the line segments to the remaining connected runs of hits using a recursive descent algorithm reported by Popplestone and Ambler[3]. A line is constructed between the two endpoints of a connected run of hits. The position on the run where the maximum deviation from the run to the straight line is noted, and if the deviation exceeds a threshold, then the run is split at this point (denoting a new endpoint), and the algorithm is recursively applied to the run on each side of the split. The recursion terminates if either the maximum deviation does not exceed the threshold (indicating the presence of a line segment), or if the remaining run length is less than some threshold. It is realistic to employ such thresholds because they are derived from known dimensional tolerances of the task. The result is an array containing one or more segment structures. The endpoints, slope, and length of each segment are components of the structure.

Identify Joints
Given the array of segments, the next step is to identify the type of joint depicted by the segments. There are three possible joints: butts, laps, and T-joints. Gaps may be present, and the angle between the metal sheets making up the joints may vary. The array of segments is applied to a decision procedure that makes explicit some geometric and linguistic criteria that any legal joint must observe. Given a list of segments, the system first determines whether it represents a legal lap joint. If it fails the lap joint test, then the system determines whether it is a T-joint. If that test fails, then the system cannot classify the joint, so another image is captured and the procedure repeats. If four successive images cannot be classified, then the system reports an error. The need to perform the tests in this order result from two important linguistic constraints: (a) lap joints contain a small fillet-like shape, so the presence of a lap joint somewhere in the list of segments rules out the possibility that a T-joint occurs anywhere in the list; (b) the butt joint is a special case of the lap joint, so the presence of a lap or T-joint rules out the possibility that a butt joint appears. Rules for identification of each of the three types of joints have been constructed, but space does not permit a full report here. We now briefly summarise what is involved in identifying a lap joint.

Fig. 3 depicts how the light stripe reflects from a lap joint to form segments, and shows how the centre of the seam can be detected. Two representative segments, A and B, are sufficient to represent the lap joint. The right endpoint of segment A we call P1, and the left endpoint of segment B we call P2. We describe below how the system chooses the representative segments A and B given a list of segments. Segment E in Fig. 3(a) results from the edge of the top surface. For the pre-production prototype without any gap, segment E should join points P1 and P2 as shown. However, the cut edges of the metal sheets are likely to be specular reflectors less than 2mm thick, and therefore may produce a large bloom in the image where segment E should appear. To get around this problem, the system first chooses A and B as described below, and uses the distance between P1 and P2 to represent E, rather than any image information that might appear. The hit hypothesiser will fill in every pixel on the straight line between P1 and P2 only if no gap is present, which turns out to be the proper thing to do, but the information is not trusted in this case. When inspecting the pre-production assembly, the length of E is proportional to the metal thickness, and angle ϕ is used to describe the line that will need to be hypothesised when the production assembly (with possible gap) is inspected. Fig. 3(b) shows how ϕ is used for a lap with a gap.

The circles in each set of segments denote the system's estimate of where the centre of the seam is. In Fig. 3(a), the centre is the corner between the bottom sheet and the edge of the top sheet. This is point P2, the intersection of segment E and B. The same applies to Fig. 3(b), but notice the portion of segment B to the left of point P2'. This part of the segment is where the stripe is seen through the gap on the bottom surface.

The lap joint can be approached from two possible directions, and a gap may appear, giving four possible classes of joint shape. A list of segments derived from the image is defined to represent a lap joint if and only if two segments A and B of maximal length exist that satisfy the following rules:
1. A minimum length condition: the lengths of A and B exceed some threshold. This is to ensure that informative segments are used.
2. A test to determine on which side of the seam the top surface of the lap appears. This involves comparing P1 and P2, and the relative slopes of A and B in the image plane. Only a legal lap joint will have a distinguishable top surface to the right or left of the seam.
3. The absolute value of the angle between A and B is less than a threshold. This rule distinguishes laps from T-joints.
4. The vertical displacement from P1 to P2 must not exceed a threshold, but must be at least as great as the metal thickness. Note that this distance is an image row distance, which corresponds to a distance in depth along the optical axis.

Other sets of rules are used to recognise T-joints and butt joints. Once a list of segments is identified with a particular joint type, the following information is associated with the appropriate endpoint in the taught welding program: the slopes of A and B, and an estimate of where the centre of seam is. A metal thickness factor (the distance between P1 and P2) and angle ϕ are also stored for lap joints.

Computing Errors between Model and Production Assembly
During the correction phase, information collected during the teach phase (slopes of legal A and B segments, estimate of seam centre, etc) is considered as a model of what to look for when inspecting the production assembly. The system now attempts to match the model with the segments derived from an image of a joint on the production assembly. Let the following terms denote the components of the model: left.slope (taught slope of A), right.slope (taught slope of B), (cr,cc) (row and column indices of seam centre), and ϕ (observed angle between bottom sheet and edge for laps). Given a list of segments derived from the image of a joint on the production assembly, the system now attempts to find segments A and B that satisfy the following rules:
1. The absolute angle between A's slope and left.slope does not exceed a threshold, and the absolute angle between B's slope and right.slope does not exceed a threshold.
2. Compute an estimate (sr,sc) of the seam centre. Angle ϕ is used to construct a seam centre for lap joints as previously described. Compare (sr,sc) with (cr,cc) to compute gap, stand-off error, and lateral error. Fig. 4 shows how this is done for the lap joint, and similar constructions hold for the T-joint and butt joint. Gap must not exceed a threshold, and stand-off and lateral error must not

exceed a threshold. Note that gap width is defined along the horizontal image axis, indicating the projection of the gap onto a plane normal to the torch. This is an arbitrary choice. An alternative would be to define gap as the distance from the edge of the top surface to the centre of the seam, for example.

If segments cannot be found that match the model with the joint on the production assembly, then the system tries again with another image. If no match is found after four images have been examined, then the system reports an error for that joint.

A number of thresholds are used in the rule base. These thresholds make explicit certain tolerances that are observed in the production of sheet steel assemblies. Thus, the thresholds are readily derived and stable. It is not necessary to determine thresholds when teaching the robot.

Results

The system as described here has been tested on more than one-hundred different images. Workpieces with a variety of joints have been used, all constructed with thin planar sheet steel pressings from about 1 to 2mm in thickness. Nearly all of the joints have been T-joints and lap joints, reflecting the prevalence of their use in the motor car industry. Stand-off errors and lateral errors of ± 3mm are regularly corrected to within ± 0.3mm, and about 20% of the cases are corrected to within ± 0.1mm, which is the resolution limit of the system. These are safe margins of error, well under the target tolerance of 0.5mm. The system has corrected some gross errors in jigging of ± 3 to 5mm, although it would not normally be expected to do this. The system has been used in close proximity (a few cm) to the arc with no problems so far. Although it has been used to correct endpoints, it has not been used to track seams in real-time because of computational limitations. The system has also been used in an informal "demonstration" mode, in which it tracks a small workpiece held in a person's hand (with a servo lag time of 2 seconds due to image processing time).

Gaps ranging from 0 to 2mm have been successfully measured to within ±0.3mm. However, about 30% of the gap width estimates exceed the true gap by more than 0.3mm. These cases are often wildly in error, with gap widths of 2mm being reported for true gaps of 0.5mm. This is due to an unsophisticated and incorrect treatment of segment endpoint location near the seam, and will be rectified in the next version of the system. The increased computational effort required by the new version will not be noticeable.

The total time required to process an image and to compute the joint type and error terms is about 1.75 seconds. Of this, 99% of the time is spent convolving the image with the discrete DOG mask to construct the hit list. We are currently developing a circuit to construct the hit list at video frame rates.

References

[1] Clocksin, W F, Barratt, J W, Davey, P G, Morgan, C G, and Vidler, A R. Visually guided robot arc-welding of thin sheet steel pressings, *Proc. 12th Int. Symp. Industrial Robots,* Paris, IFS (Publications) (June 1982).

[2] Agin, G J, and Binford, T O, Computer description of curved objects, *Proc. 3rd Int. Conf. Artificial Intelligence,* Stanford, USA (1973).

[3] Popplestone, R J, and Ambler, A P, Forming body panels from range data, Research Report 46, Department of Artificial Intelligence, University of Edinburgh (1977).

[4] Hill, J, and Park, W T. Real-Time control of a robot with a mobile camera, *Proc. 9th Int. Symp. Industrial Robots,* Washington D.C., Society Manufacturing Engineers, (March 1979).

[5] Hunter, J J, Bryce, G W, and Doherty, J. On-line control of the arc welding process, *Proc. Int. Conf. Developments in Mechanised, Automated, and Robotic Welding,* London, Welding Institute (November 1980).

[6] Vidler, A R. Equations relating welding parameters, Robot Welding Project Report, (in preparation) (1982).

[7] Jensen, K and Wirth, N. PASCAL User Manual and Report, *Lecture Notes in Computer Science 18,* Springer-Verlag (1974).

[8] Marr, D, and Hildreth, E. Theory of edge detection, *Proc. of the Royal Society (B)* Vol. 207, pp. 187-217 (1980).

Chapter 5
DEVELOPMENTS –
ASSEMBLY/PART PRESENTATION

Robot assembly is still at an early stage of application and vision is seen as a key area in need of development. The four papers reflect the subjects now being researched; these are robot guidance, assembly process control and part presentation.

INTELLIGENT ASSEMBLY ROBOT

Michinaga Kohno, Hiroshi Horino and Mitsunobu Isobe, Production Engineering Research Laboratory, Hitachi Ltd., Japan

Reprinted from Hitachi Review, Vol. 30 (1981), No. 4, pp. 211–216, Hitachi Ltd., Japan

Abstract

An experimental intelligent assembly robot comprising two articulated robots and a vision system has been developed. The pilot system is capable of visual recognition of location and orientation of parts which are randomly placed, and of performing coordinated motions of two arms. A new robot mechanism in which various configuration of joints can be built from a basic model with six degrees of freedom has been developed. In order to carry out such advanced control of the robot as vision interaction and multi-robot coordination, a control technique employing hierarchical structure of two microprocessors has been applied. A new robot language has been developed for easy description of robot programs involving such advanced control. A vision system using a partial pattern matching technique has proven to be well suited for robot vision.

Introduction

In accordance with the need for flexible automation of assembly processes, a pilot robot-vision system, which contains elements of essential technologies fundamental for the future research and development of assembly robot systems, has been developed.

The experimental system is composed of a pair of articulated robots and a vision interaction system, and its outlines and features, which are shown in Fig. 1, are:
1. A new configuration of an articulated robot mechanism with a modular structure.
2. Coordinated operations of the two robots.
3. Robot motion generation based on visual information.
4. A new robot language for ease of programming the robots.
5. A hierarchical structure of control employing a multi-processor configuration to implement the above-stated features.

The system was first publicly shown at the technical exhibition held in Tokyo in November 1980 to celebrate Hitachi's 70th anniversary. A demonstrational operation of assembling toy trains, as shown in Fig. 2, was performed during the exhibition.

Robot mechanism

The two robots constituting the pilot system are identical in their structures and specifications, and the description in this section is of a single robot. The robot has an articulated structure with dimensions similar to those of a human arm as shown in Fig. 3(a). On the basis that the robot is intended for light-assembly applications, the load capacity was given as 3kg, and DC servomotors are used to drive the mechanism as they were envisaged as the most suitable means to drive such a light load. The specifications of the robot are shown in Table 1.

The robot has six degrees of freedom (DOF). In most conventional applications of the robots such as arc-welding and transfer operations, five DOF are sufficient because the robot only repeats the programmed motions which have already been proven possible for the robot through teaching. However, six DOF are needed for an intelligent robot, whose motion is not always programmed in advance but is created from circumstances. The reason for this is that the robot has to have independent freedom of motions corresponding to six parameters that are necessary to define the position and orientation of an object in three-dimensional space in order to obtain arbitrary position and orientation of its end-effector within its operational range in accordance with information from its vision system. Based on these considerations, the intelligent assembly robot was designed to have six DOF.

After the comparison of several feasible configurations of the the joints, a configuration with twist of forearm, shown in Fig. 3(b), was chosen for the following reason. The axes of forearm twist, wrist bend and wrist rotation intersect at one point and therefore these joints as a whole are kinematically equivalent to a spherical joint with three DOF. This equivalence effectively simplifies the

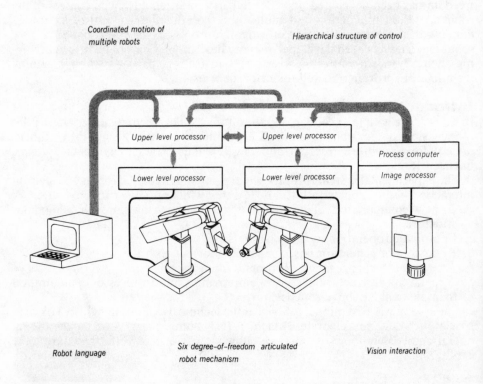

Fig. 1. System Configuration and Features of Intelligent Assembly Robot.

Fig. 2. Intelligent Assembly Robot at Hitachi Technical Exhibition.

coordinate transformation in the robot control algorithms and thus reduces computation time.

While the robot has six DOF, surveys of possible applications of robots to manufacture in practice show that many operations can be performed by robots with fewer degrees of freedom. The use of six DOF robots for such simple operations can not be economically justified. To satisfy this requirement and create a structure of justification, a concept of modular-structured robots has been developed for some types of robots that are widely used in industry. However, due to the essential difference in structure, the modular robots have been limited to robots of cartesian, cylindrical and spherical coordinates. A new structure of an articulated robot, which allows for the choice of various freedom configurations, was developed and applied to the intelligent assembly robot. Here, the six DOF model is regarded as the basic model and various models with fewer DOF can be built through modifications of the basic model. A five DOF robot mechanism with the orientation of its wrist fixed, for example, can easily be built simply by removing the motor for wrist bend and by tying up the rear end of the drive shaft mechanism to the rotational column.

With a line-up of variations ranging from a simple pick-and-place manipulator to a very versatile six DOF model, a wide range of operations can be economically robotised by using robots each being suited to the requirements of its jobs.

Control

The control of articulated robots requires more complicated procedures than that of cylindrical or cartesian coordinate robots. Even in a straight line motion of the arm tip, rotational motions of the joints are complex. To control such motions precisely, the following processes have to be performed in a limited duration of sampling cycle time:

Table 1. Specifications of Robot Mechanism

Configuration	Articulated
Degree of freedom	6
Load capacity	3kg (incl. tool weight)
Static tip force	10kg
Maximum tip velocity	1.2 m/s
Repeatability	±0.1mm
Drive	DC servomotor

1. The interpolation between the present position of the arm tip and its goal position through a predetermined trajectory.
2. The coordinate transformation of the position of the arm tip into angular displacements of each joint.
3. The rotational velocity command calculation from angular displacements of each axis given by the software servo to servomechanism.

In addition to the above-mentioned control requirements, the intelligent assembly robot has such functions as pattern processing for motion generation based on sensory information and communication between robot controllers for coordination of multiple-robot motions. Further, since the sampling cycle time, during which all of these processes have to be carried out, directly affects the accuracy of the robot, time required for these processes must be minimised. In order to carry out such advanced control of the robot, a new control technique using two microprocessors for each robot has been developed.

The complex tasks stated above have been split and allocated between two HMCS 6800 microprocessors, which are hierarchically linked together. The upper level processor is responsible for the supervision of the whole system, processing of the data obtained through teaching, calculations of interpolation and coordinate transformation, and communication with processors for visual recognition and with the controller of the partner robot. The lower level processor is responsible for the servo control of the joints. The upper and lower level processors are linked by means of a dual-port random access memory (RAM). The hardware structure of the robot control is given in Fig. 4.

The lower level processor, which is solely dedicated to the software servo, has fewer tasks than the upper level processor. Therefore, a split servo technique was introduced to the lower level processor, so the lower level processor performs the

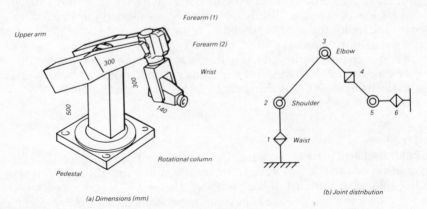

(a) Dimensions (mm)

(b) Joint distribution

Fig. 3. Dimensions and Joint Distribution of the Robot.

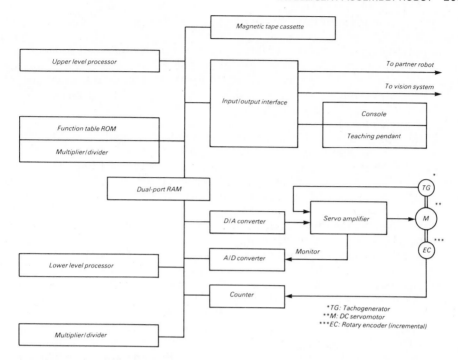

Fig. 4. Block Diagram of Robot Control.

servo control procedures twice per sampling cycle of the upper level processor. Consequently, the sampling cycle time of the lower level processor was greatly shortened and its servo characteristics were improved.

As a result of the hierarchical structure of the two processors and the task allotment, the sampling cycle time of the whole system and of the servo control of the joints were curtailed to 35.2 ms and 17.6 ms, respectively. These values indicate that, despite the addition of advanced functions like vision interaction and two-arm coordination, the control procedures of this newly developed system are as fast as conventional robots, where the processors are mostly dedicated for the servo control. Accordingly, it can be concluded that the control system discussed above is suitable for the control of the intelligent assembly robot.

Robot language

The programming of conventional robots of the playback type has been performed mostly by guiding the robot through its motions by using a manual operation box, called a teaching pendant. This method is practical only when all the details of the motions are known at the time of programming. However, in the application environments of robots involving decision-making based on sensory information and synchronisation with peripheral equipment and/or other robots, programming of the robots can not be made only by teaching but some form of written programming is needed.

Upon these background conditions, a prototype robot language has been specially developed for this pilot robot system. The commands of the language are functionally classified into the following three categories: teaching command, program command, and monitor command. The configuration of the software system including the robot language is shown in Fig. 5.

The teaching commands are used for manual operation of the robot and for input and modification of location data of the robot. These commands are input to the controller by pressing the buttons on the teaching pendant. Four different modes of manual operations have been made available and can be selected in any combination during teaching. These modes are:
1. Independent motion of joints.
2. Motion of the arm and wrist in cartesian coordinates fixed to the ground.
3. Motion of the arm in cartesian coordinates fixed to the wrist.
4. Motion of the arm and wrist in arbitrarily shifted cartesian coordinates.

These manual operation modes provide human operators with robot motions matching the human faculty of sensation, and the labour for teaching has thus been greatly reduced.

The program commands are used for the description of motion sequence of the robot, and are input through the control console. There are four categories of program commands, which form the main structure of the robot program, and have the following functional features:
1. *Motion.* Several motion commands were made available such as linear interpolation and simultaneous control of the six axes. A sequence of robot motion can be described by arbitrary combination of these motion commands, and a rich vocabulary of motion commands allows for easy description.
2. *Vision interaction and triggering.* With the vision interaction function and with the triggering function for peripheral equipment, the generation of location data based on the visual information and synchronisation with peripherals machines can now be easily programmed.
3. *Program flow control.* The program flow control commands have been elaborated to include synchronisation of multiple robots, motion repetition, conditional branching and dwelling. Thus it is easy to describe complicated operation flows.
4. *Arithmetic logic.* The arithmetic calculation commands for location data were implemented and the shift of location variables in a program was thus made possible. This is very useful in some applications like palletising, where the same operations are repeated at different locations with a fixed pitch.

The monitor commands are used for the selection of task control, and include the following:
1. Editing and modification of the robot motion proram.
2. Selection of teaching and execution modes.
3. Storing and loading of program and location data to and from magnetic tape cassettes.
4. Display of program and location data onto terminal devices.
5. Recover from errors.

In the editing, modification, teaching and execution modes, partial execution, step execution and alternating processing among these modes are possible. This feature greatly facilitates the debugging of programs.

Among the functions of the language stated above, the vision interaction and multi-robot coordination are particularly distinguished. By simple combinations of visual information input commands and conditional branching commands, such complicated operations as follows can be easily programmed:
1. Recognition of location and orientation of parts which are randomly placed.
2. Selection of assembly procedures depending on location, orientation or manners of lying of parts.

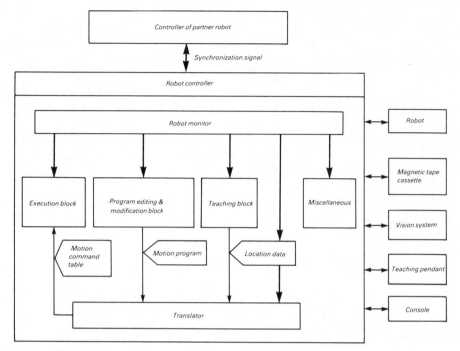

Fig. 5. Structure of Software System.

3. Compensation of positional error of robot-hand-held parts for accurate mating with other parts.

The description of multi-robot coordination, on the other hand, has been made simple by developing a language structure, with which motion flow involving multiple robots can be written in a single program. Through translation procedures, this source program is automatically divided into programs for individual robots, and the synchronisation commands between the robots are also automatically inserted as required in each program.

Vision

Emphasis of the vision system used in the pilot system was placed on demonstrating the feasibility of using vision which is fast, reliable and reasonably economical as a robot vision. For quick processing, partial pattern matching technique was employed, and because of the experimental nature of the system, circular targets are marked on the parts.

Fig. 6 shows the structure of the vision system. A HIDIC-80 process computer with 32k word core memory controls the image processor, executes various recognition procedures, and communicates with the robot controllers through interfaces. The TV cameras are mounted on a support frame over the operational area of the robots and look downward to obtain two-dimensional images of the parts placed below. The image processor is dedicated hardware for fast image processing and is composed of five processing units linked by an image data bus. The analogue processing unit takes in image signals from TV cameras and outputs them to the image data bus through A/D converters and binary conversion circuitry. The threshold level of the binary conversion circuitry can be set by the computer.

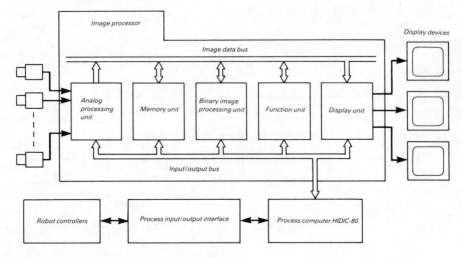

Fig. 6. Configuration of Vision System.

The memory unit is used for temporary storage of image data. The binary image processing unit is equipped with four counters and 16 × 16-pixel partial pattern matching circuitry for binary image. The inclusion of the counters makes it possible to take various measurements such as area and width of patterns. The function unit performs summation, multiplication, logic operations and integration of image data. The display unit drives the display devices to show original analogue images, binary images and images stored in the memory.

In the demonstrational operation of assembling toy trains at the technical exhibition, the intelligent assembly robot visually recognised the location and orientation of the base and main body of the locomotive although they were placed randomly but separately. The targets were placed on the top and on both flanks of the main body and on the top of the base. Each face had three targets, that were positioned in different manners depending on the face. The location, orientation and manner of lying, upright or sidelong of the part are thus calculated from the relations among these three targets. To avoid mal-recognition which could occur in such cases where the target of another part appears in the search region, the distance between the targets and the presence of parts between the targets are checked.

Visual compensation of positional errors, which could be caused by the change of holding manners and by the limited resolution of the above-mentioned recognition procedures, can be performed by recognising the location of a target on a part held by the robot in a close-up view of the vision system. The discrepancy of the target location against the datum location stored during teaching was measured prior to placing it onto the base, and the amount of discrepancy was transmitted to the robot controller for error compensation.

Conclusions

An experimental intelligent assembly robot comprising two articulated robots and a vision system has been described. The pilot system has formed a technological foundation for the future research and development of assembly robots and relevant technologies upon the experiences of previous robots. Further efforts are being made towards more practical implementation of each branch of the technology, namely, robot mechanism, control, software and vision.

SIMPLE ASSEMBLY UNDER VISUAL CONTROL

P. Saraga and B. M. Jones, Philips Research Laboratories, UK

Reprinted from Digital Systems for Industrial Automation, Vol. 1, No. 1, pp. 79–100 (1981) by kind permission of Crane, Russak & Co. Inc., USA

Abstract

Until recently, industrial assembly has been performed either by fixed automation or by people. Fixed automation systems are in general fast, reliable, and appropriate to mass production, while manual assembly is slower but more adaptable to change. The purpose of 'flexible automation' is to provide an alternative to these two existing methods.

Flexible machines are intended to be modular, easily programmed to do a variety of tasks, and equipped with sensors, such as TV cameras, to observe and react to changes in their environment. The machine interprets the visual information it receives in terms of its 'model of the world' and instructs the mechanical actuators to make the appropriate action. In addition the results of the mechanical actions can be observed in order to improve accuracy and correct for errors.

This paper discusses some of the problems involved in constructing and using such a visually controlled machine for 3D assembly. These include manipulator control, lighting, picture acquisition, extracting three-dimensional information from two-dimensional pictures, machine calibration, system structure, and the use of multiple processors.

These problems are examined in the context of a particular class of assembly tasks, namely the vertical insertion of objects into fixtures. In particular an experimental system to perform a simple task is described in detail.

The experimental system

The task

The task chosen was the placement of three different size rings (each 2mm thick with outer diameters of 10, 18 and 25mm respectively) onto a tower consisting of three eccentric discs. Each disc is 0.2mm smaller in diameter than the internal diameter of the appropriate ring. In order to simulate a typical industrial situation in which the mounting point is difficult to see and is part of a larger body, the three storeys of the tower are painted matt black and are mounted on a matt black base. This means that the storeys cannot be seen from above but must be located from the side. The system design should be such that the tower can be positioned anywhere within an area of approximately 70 × 70mm. The rings are assumed to lie on a horizontal surface each within an area of approximately 40 × 40mm.

Although this is not a real industrial task, it contains the key elements of a

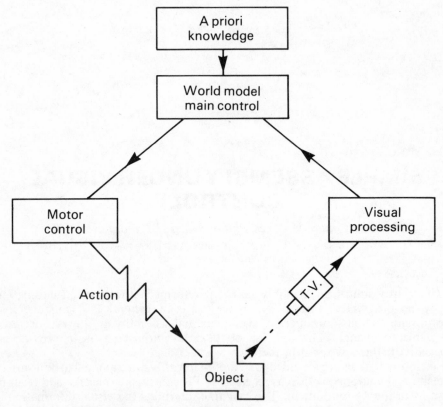

Fig. 1. Conceptual structure of a visually controlled machine.

number of practical industrial problems. The task was solved by an experimental system consisting of a computer controlled manipulator equipped with TV cameras. (See Fig 2).

The manipulator
The manipulator has four degrees of freedom of which three were used for this task. Radial, rotation and vertical (R, θ, Z) motions are provided by hydraulic actuators with integral measurement systems. Each actuator can move 100mm in 4000 steps of 25 microns each. Although the resolution is 25μ the absolute accuracy of each axis is only 0.1mm. The R and Z axes are driven directly while the θ axis has a mechanical advantage of approximately four, giving an absolute accuracy of ±0.4mm. It can be seen that by itself this manipulator is not accurate enough to perform the task.

The fourth degree of freedom which was not needed for this task is an electrically powered wrist mounted just above the gripper. The manipulator is controlled by special purpose electronics connected to a small computer (a Philips P851).

The Vision System
The size and positions of the three rings are determined using a vertical TV camera, A, as illustrated in Fig. 2. The position of the appropriate storey of the tower is located by two horizontal TV cameras (B and C). Each of the two cameras sees a projected image of the tower, which has been back illuminated by light parallel to

the optic axis of the TV cameras. The shape of the tower in these parallel projections is independent of the position of the tower in space and hence can be easily analysed to give the position of the tower in the TV image. The two TV positions can then be combined by conventional triangulation to give the position of the tower in space. This method has enabled the potentially complex task of the 3D location of the tower to be divided into two simple 2D tasks.

The optical system for parallel projection, the visual processing for ring and tower location, and the combination of two views to locate an object in space are all described in more detail later.

The Computer System

The overall structure of the system is shown in Fig. 3. It consists of two computers interconnected by slow V24 links, one link being dedicated to each direction. The TV cameras are connected to the Philips P857 by a flexible interface TELPIN, developed as part of the same project which can handle up to eight different channels selectable by the program. The interface can provide analogue or digital information from the input channel and can select any area from the TV picture and one of four resolution values. The threshold value used to obtain binary pictures can also be computer controlled. The P857 performs the overall strategic control of the machine and the processing of the pictures to extract the positional information needed for the task. The smaller P851 carries out the detailed path control of the manipulator and some higher level functions involving several movements and gripper changes. Thus, once the ring has been located the P857

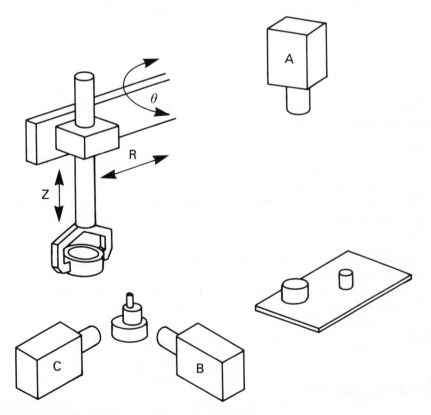

Fig. 2. Experimental system for assembling rings onto a tower.

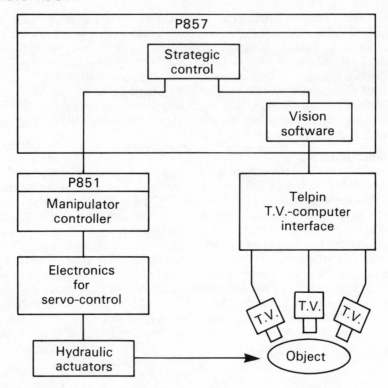

Fig. 3. System structure.

issues a single command to the P851 to pick up the ring and bring it over the tower. The P851 then carries out the detailed moves and grip changes independently, and informs the P857 that it has completed the action.

System Operation

The flow chart of the system operation is shown in Fig. 4. Once a ring has been located, and its size determined, the manipulator is instructed to pick up the ring and take it to a position high above the approximate position of the tower. This operation is under the control of the P851, and the view of the tower is not obscured by the manipulator during this period. Therefore the P857 can use the two horizontal views to determine the position in space of the appropriate storey of the tower at the same time as the mechanical operation is taking place.

If the tower storey is successfully located then the manipulator lowers the ring to a position immediately above the appropriate storey of the tower, where the relative position of ring and tower can be checked using the horizontal V cameras. The manipulator now moves the ring until it is exactly above the correct storey. The ring is then placed on the tower.

While the P851 is controlling the final placement of the ring onto the tower, the P857 is using the vertical TV camera to locate the next ring.

Picture processing

In order that the assembly task can be carried out at a realistic rate it is necessary that the picture processing employed be fast. Thus very simple algorithms have to

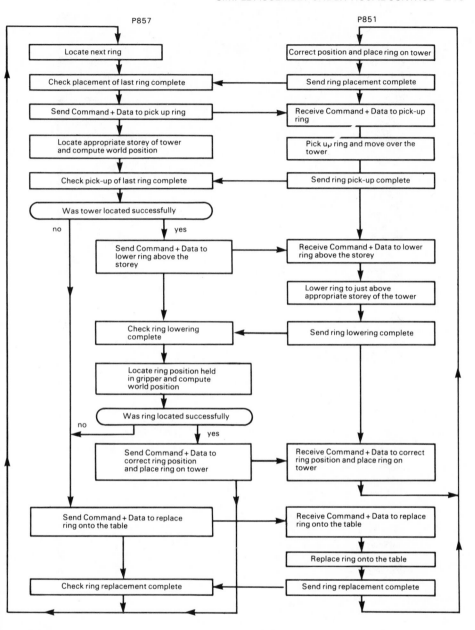

Fig. 4. System operation flow chart.

be employed. To illustrate this the optical processing used in the task will be described in more detail.

Ring Location

One of the three prescribed areas of the field of view (Fig. 5a) of the vertical camera is sampled at a low resolution into the P857 as a five bit grey level image. The grey level image is thresholded and the binary image is edge traced,[2] and black areas are

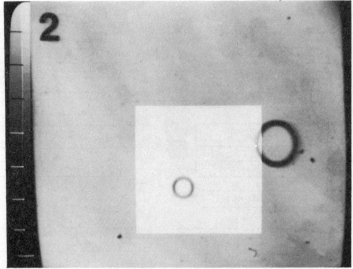

(a) Low resolution: Determination to ring size and approximate position.

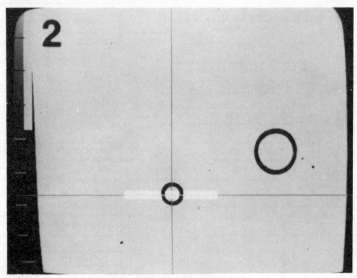

(b) High resolution: Determination of X coordinate.

Fig. 5. Ring location – Area 2.

located. If an object corresponding to one of the possible rings is not found, furthe attempts to detect a ring by changing the threshold are made. If a ring still cannot be found the next area is examined. Once a ring is detected its approximate centre is calculated and two further regions centred on the approximate ring position are sampled at high resolution to determine the accurate X and Y coordinates of the ring centres in TV units (Figs. 5b, 5c). The regions are searched from both ends until a white black edge is found and the appropriate coordinate is taken as the mean of these two edges.

The position of the ring is then converted to world coordinates and the

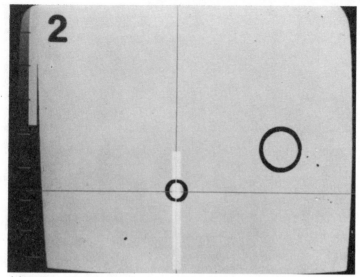

(c) High resolution: Determination of Y coordinate.

Fig. 5 (contd.). Ring location – Area 2

manipulator is instructed to pick up the ring and bring it to a position approximately above the tower.

Location of Tower

The same processing is carried out on each of the two horizontal TV channels used to locate the tower. First a large coarse resolution scan (Fig. 6a) is carried out to locate a black horizontal bar of the correct size for the approximate storey. Then a second smaller scan (Fig. 6b) at higher resolution is taken at about the approximate centre of the tower. This is used to locate the top of the storey.

Small scans at high resolution are then taken at each side of the tower (Figs. 6c, 6d) and the black/white boundary points found. The separation between these points is computed and only those pairs whose separation is within five pixels of the correct storey widths (which range from 50 to 110 pixels) are accepted. When eight pairs have been found meeting these criteria, the mean of their centre position is found and taken as the centre of the storey. The mean of the error in width between the acceptable pairs and the correct storey width is used as an error function to correct either the expected width of the tower or the threshold for the next cycle of the system. Thus the system is self-correcting and can allow for drift in the TV video level. The process is repeated on the second camera. The TV coordinates are converted to a line in world space parallel to each optic axis and then the point of closest approach between the two lines is found and taken as the world coordinates of the tower. The tower may be placed in any position where all storeys can be seen by both cameras. The field of view of each camera is set to be approximately 100 × 100mm which completely covers the 70 × 70mm area in which the tower may be.

Visual Feedback

The sampling system can resolve approximately 600 × 600 points in the TV field of 10mm × 100mm giving a resolution of ≈ 0.17mm. The TV position of the tower is known to a rather higher accuracy since it is obtained by averaging a number of edge points. The accuracy of the world position depends both on the accuracy of

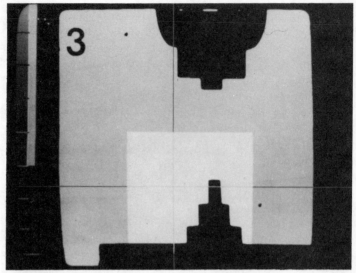

(a) Low resolution: Initial search.

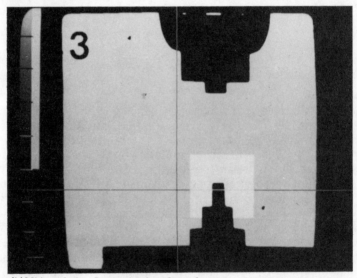

(b) Mid resolution: Determination of top of storey.

Fig. 6. Tower location on TV channel 3.

the TV position and on the accuracy of calibration of the camera. Experimenta
results give an accuracy in the world position of roughly ±0.1mm.

However, as mentioned earlier, the clearance between the rings and the tower i:
only 0.1mm in radius, and while the manipulator has a repeatability of 0.025mm ir
R and Z and 0.1mm in θ, the non-linearity in the measuring system and rounding
errors in the geometric modelling of the manipulator result in an absolute accuracy
of only 0.1mm in R and Z and 0.4mm in θ. These figures, together with the
inaccuracy in the location of the tower, will clearly not allow the ring to be safely
put on the tower.

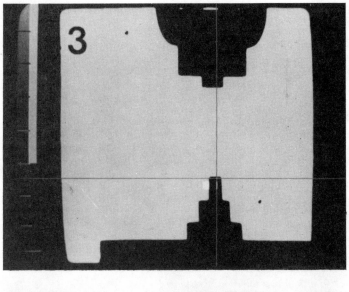

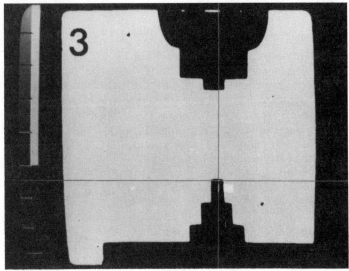

(c)
(d) Final position determination at high resolution.

Fig. 6 (cont'd). Tower location on TV channel 3.

This problem may be overcome by using visual feedback, that is, by lowering the ring to a position immediately above the appropriate storey of the tower and then observing the relative displacement of ring and tower in the two horizontal cameras. A narrow high resolution scan (Fig. 7) is made to locate the ring projecting below the gripper. The location of the ring depends again on finding a pair of white/black and black/white vertical edges with the required separation, and the same updating mechanism is used to cope with apparent changes in ring size due to camera drift. The world displacement between ring and tower may now be

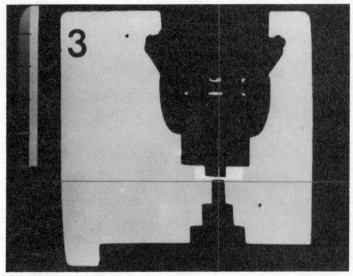

Fig. 7. Checking ring position over tower on TV channel 3.

calculated and the manipulator ordered to make the appropriate adjustment before placing the ring on the tower.

The use of this visual feedback has proved critical in the successful implementation of the task. It could not have been achieved with a fixed vertical camera since the manipulator obscures the critical area which has to be observed.

The visual feedback also allows the position of the ring in the gripper to be checked. If the ring is not square in the gripper then it will not be recognised and the program will instruct the manipulator to replace the ring on the table. Thus the critical region of the task is protected by a visual interlock.

The parallel projection optical system

As previoulsy mentioned, the two horizontal cameras use parallel projection optics. The system used[3] is based on that employed in the conventional tool room projector.[4] The optical arrangement is as shown in Fig. 8. The object is back illuminated by a parallel beam generated by a small light source, S, at the focus of a fresnel lens, L1. A combination of a second fresnel lens (L2) and a camera lens (L3) produces a real image of the parallel projection of the object on the face of a TV camera tube, C. It will be seen that this method of viewing the object has many advantages, when applied to the determination of the position and orientation of objects.

Using this optical system, the image size is essentially independent of the position of the object. This considerably simplifies the picture analysis which can remain constant even when the object is moved nearer or further from the camera. The position of the object perpendicular to the axis is of course given directly by the position of the object in the TV image.

The optical arrangement is virtually immune to changes in ambient illumination since only light which is nearly parallel to the optic axis passes through the camera lens. The smaller the aperture of the camera lens, the higher the immunity and, since the TV camera needs relatively little light, this aperture can be made very small. This feature may be very useful in a factory environment where ambient

ighting can be difficult to control. This very small aperture also gives a large depth
of focus.

Finally, it is easy to calibrate these systems and to combine them to produce
object location systems. Both these features are discussed in more detail below.

Object location using parallel optics

Because the optical systems use parallel light and are immune to ambient illumina-
ion it is generally possible to have a number of systems at various positions within
a machine without the systems interfering with each other. Combination of systems
is usually necessary to locate objects in space. A single axis system can only 'locate'
an object point as lying on a line in space parallel to the optic axis. Unless the object
is otherwise constrained (e.g., the camera is vertical and the object is lying on a
horizontal plane[1]) two systems are required to locate an object in space.

The location of an object in space from two views requires that the projection of a
common object point be identified in each image. It is only then that the object
point can be defined by the intersection of two projecting rays.

In the case of the tower, each storey is a cylinder. Such an object viewed from two
calibrated horizontal cameras is illustrated in Fig. 9. The common point, the centre
of the top surface of the cylinder, is projected as the centre of the black rectangle
appearing in each image. The direction of the two cameras is not critical; the more
orthogonal they are, the better the accuracy of the location.

It is interesting to consider the extension of this method to other $2\frac{1}{2}$D objects, by
which is meant objects with vertical parallel sides whose cross section is not
circularly symmetric. The projected profile seen by each of the cameras will be a
rectangular black area as in Fig. 9, but the width of the profile will depend on the
orientation of the object to the optic axis of the camera. The problem of determining a common point in both views generally becomes more complicated, although
for some regular shapes (which often occur in industry) such as a square, rectangle,
or hexagon, the centre of the profile still corresponds to the projection of the centre
of the object.

Because the object is not circularly symmetric it is probably necessary to
determine the orientation as well as the position. Back lit parallel projection has an
inherent 180° ambiguity since there is no difference in profile if the camera and
light source are interchanged. In many industrial applications this ambiguity may

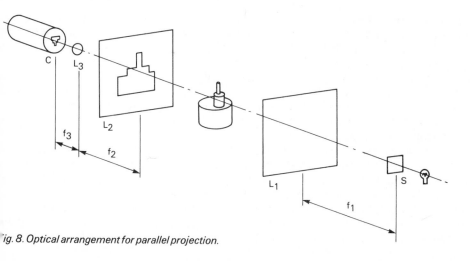

Fig. 8. Optical arrangement for parallel projection.

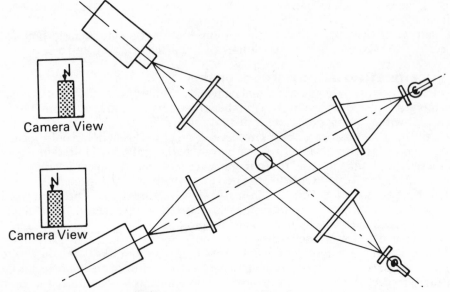

Fig. 9. Location of an object from two views.

not be a problem, either because the assembly possesses twofold symmetry, or more generally because the crude orientation of the object is known, possibly from some previous stage of assembly. A situation of this type might occur in the final stages of assembly when the object is about to be inserted into the fixture. In this case a camera could be arranged to be at an angle where the rate of change of profile width is large.

If the approximate orientation is not known it still may be possible to remove the ambiguity using basically the same optical system. If lights are shone from the direction of the camera towards the object, then only light reflected parallel to the axis will be accepted by the camera. This results in a grey level image of the object, again of constant size, in which details of the front surface of the object can be seen. By observing this image and determining only the presence or absence of some feature it should be possible to remove the 180° ambiguity. A single measurement of the profile width will not be sufficient to determine the orientation, but using two or more cameras to view the object will in general allow the object's orientation to be determined. Alternatively, the object could be rotated by a known amount in the field of view of one camera.

An object's profile width may be determined analytically if the cross section has a simple form. Alternatively the profile may be determined experimentally and stored in a reference table in a training phase. Then, in the operating phase, the width of the profile can be measured and looked up in the table to find the corresponding value of the orientation, θ.

If the object is such that it is not easy to determine the position of a known object point in the profile, then a similar method involving the use of a table to determine both the orientation and the position may be used. For example, the position within the profile into which a particular object point, P, is projected when the projection angle is θ, can be stored in the same table as the width during the training phase. Thus in the operating phase, once θ has been found, the position of P in each projection is given by the table. The position of P in space can now be obtained by

(a) Pointer in manipulator gripper.

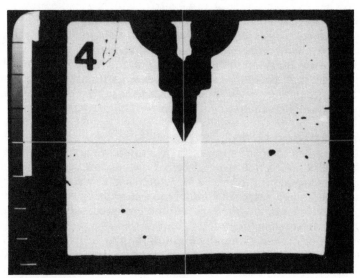

(b) TV image of 4 (a) location of pointer tip.

Fig. 10. Calibrating TV channel 4.

converting the TV coordinates to world coordinates and calculating the intersection of the two rays.

More complex fully three-dimensional shapes may often contain an element of simple cross section which can be used to solve the 'correspondence problem'. Because the image size is constant the position of this element within the complex outline can be found by simple dead reckoning from the bottom of the object.

Calibration

In order to use a visually controlled machine it is necessary to relate the images from the various TV cameras to the movements of the manipulator and to the positions of the various parts being handled. The calibration process determines the relationship between 'frames of reference' associated with each part of the overall system. Whenever the machine is modified or adjusted, the relationships change, and recalibration is required. If effective flexibility is to be achieved it is important that calibration is both simple and rapid.

It is generally most convenient to relate all the manipulator and TV frames to a single arbitrary Cartesian 'world frame'. Thus, if any unit is moved, only its relationship to the world is changed.

Calibration of the manipulator involves determining the relationship between the movements of the R, θ, Z actuators and the world frame. It can be determined from the geometrical structure of the manipulator and the 'extension per step' profile of each actuator. Since the three manipulator axes being used for this task are decoupled, the geometric relationship is not complicated.

In general, calibrating a TV camera means determining where it is in 3D space and in which direction it is pointing.[5] Using the parallel optics system, it is not necessary to know where the camera is, but simply to know in which direction it is pointing relative to some 'world coordinates' and also the magnification of the lens. Suppose the optical system is being used in conjunction with a manipulator whose movements are known in world coordinates. Then if the manipulator is moved through known distances in defined directions and the results of those motions observed in the TV image, the 4 × 4 matrix, which specifies the relationship between the TV and the world frame, may easily be calculated.

In our experimental system we place a rod with a sharp point in the jaws of the gripper (Fig. 10a). Since the image of the rod remains constant in size as the manipulator moves, the point of the rod is very easy to recognise (Fig. 10b), and the movement of the manipulator is easy to observe. This simple procedure allows cameras to be placed in positions and angles which suit the task. Calibration for a new position requires less than two minutes.

Distributed processing

In the example given of positioning rings on a tower we used two processors. The division of the task between the two processors is such that only simple commands with parameters are passed to the P851, which acts as a slave of the P857. This simple division allows a degree of parallel processing and a consequent increase in overall speed of the system. There is no problem of data division in this system since all the data is stored in the P857.

It is possible to devise other organisations using more processors to allow additional processes to be carried out in parallel. For example, if a separate processor was used for each horizontal camera then the determination of the tower and ring position in each view could be carried out simultaneously.

It is necessary to consider other factors before adding processors indiscriminately. For example, the division of tasks must not require large amounts of data to be transferred between processors, nor should it require the maintenance of the same data in two processors. The distribution of processes among processors should also reflect the conceptual structure of the overall system in order to produce a modular configuration which is easy to understand and modify. Another important consideration is the relative speeds of the various parts of the system. There is obviously no point in adding processors to speed up the location of the tower if, as is already the case, the tower is located in approximately the same time as it takes the manipulator to move the ring to above the tower. There are three speed determining factors in a visually controlled machine with multiple processors. They are: the speed of the manipulator, the speed of the visual processing, and the speed of intercommunication between the processors. Changing any one of these factors can completely alter the optimal number of functions to be performed in each processor.

Conclusions

The experiments described in this paper have shown that back lit profile images using parallel light can be usefully applied to some tasks in mechanical assembly. It has been demonstrated that the advantages of this approach include: rapid size independent picture analysis allowing identity, position, and orientation to be determined by direct measurements; immunity to ambient illumination; and simple 3D object location using two or more views.

Although the system has only been applied to a simple example we intend to apply the same methods to a number of practical examples. The use of many processors including specialised high-speed picture processors in systems of this type is now becoming practical. It is therefore becoming important to find methods for both constructing and using these systems to give the desired qualities of speed, modularity, flexibility, and conceptual simplicity.

Acknowledgements

We would like to acknowledge the contribution of D. Paterson and A. R. Turner-Smith who, with our colleagues at Philips Research Laboratories, Eindhoven, constructed the manipulator. We would also like to thank our Group Leader, J. A. Weaver, for his help and encouragement.

References

[1] Saraga, P., and Skoyles, D. R., 'An experimental visually controlled pick and place machine for industry', 3rd International Joint Conference on Pattern Recognition, Coronado, California (November 1976).

[2] Sarage, P. and Wavish, P. R., 'Edge tracing in binary arrays', Machine Perception of Patterns and Pictures, Institute of Physics (Lond) Conference ser. No. 13 (1972).

[3] British Patent Application No. 7942952: An object measuring arrangement, December 1979.

[4] Habell, K. J. and Cox, A., *Engineering Optics,* Pitman (1953).

[5] Duda, R. O. and Hart, P. E., *Pattern classification and scene analysis,* Wiley Interscience (1973).

A ROBOT SYSTEM WHICH ACQUIRES CYLINDRICAL WORKPIECES FROM BINS*

Robert B. Kelley, John R. Birk, Henrique A. S. Martins and Richard Tella,
Department of Electrical Engineering, University of Rhode Island, USA.

Copyright 1982 IEEE. Reprinted with permission from IEEE Transactions on Systems, Man and Cybernetics, vol. SMC-12, no. 2, pp. 204-213 (March/April 1982).

Abstract

The feasibility of robots employing vision to acquire randomly oriented cylinders has been demonstrated for the first time. An experimental robot system using vision and a parallel jaw gripper was able to acquire randomly oriented cylindrical workpieces piled in bins. Binary image analysis was adequate to guide the gripper into the multilayered piles. Complementary information was provided by simple sensors on the gripper. Experiments were performed using titanium cylinders 6cm × 15cm diameter and 7.6cm × 3cm diameter. Cycle times to acquire a cylinder and deliver it to a receiving chute ranged from 8 to 10s when a single supply of one-size cylinders was used. By using a dual supply bin configuration with one bin of each size and overlapping arm motion and image analysis tasks, the cycle times for one cylinder from alternate bins ranged from 5.5 to 7.5s per piece. In the future robots with such capabilities can be applied to enhance automation applications, especially in small batch production.

Introduction

In many manufacturing environments workpieces are supplied in bins. It is a common industrial problem to load machines with such pieces. Often human labour is employed to load an otherwise automatic machine with workpieces. Such jobs are monotonous and do not enrich human life. Young people today have increasingly higher aspirations for good jobs. While the cost of labour is increasing, human performance remains essentially constant. Furthermore, the environment in manufacturing areas is generally unhealthy and, when workpieces are inserted into machines, limbs are often exposed to danger. These factors strongly suggest that alternatives to the use of human labour be considered.

The engineering alternatives are the use of mechanical feeders, the preservation of orientation of parts throughout the manufacturing process, and the use of robots with vision to feed pieces from bins. To be viable, any approach must be competitive on an economic basis. The advantages of using robots with vision can be appreciated when the disadvantages of the alternatives are identified.

---

* This work was supported by the General Electric Company, Aircraft Engine Group.

Mechanical feeders have problems with parts jamming. This is caused typically by parts which are very thin, out of tolerance, or foreign. Some pieces have shapes which make them difficult, if not impossible, to orient. Some parts can be damaged by scratching against each other or against the orienting device. Large workpieces require very large feeders and substantial energy to be oriented. For vibratory feeders, vibrations may be conducted into other structures. Feed rates change with the number of parts in the feeder. Mechanical feeders often are excessively noisy. The cost and lead time to design, debug, support and changeover mechanical devices may be prohibitive. This is especially true for batch production applications. Even if a rejection mechanism is controlled by vision[1], most of the problems with mechanical feeders which have been cited remain.

Preserving the orienation of oriented parts is the most obvious way to avoid the problem of dealing with a bin of randomly oriented parts. This is particularly the case since commercially available industrial robots can load machines which are supplied in magazines or pallets. However the preservation of orientation is often impractical or uneconomical. When different machines work at different rates, the rate of linked machines must be set by the slowest one. Buffer storage and line splitting and merging may be necessary to prevent the failure of one machine from stopping the entire line. If a robot is used to preserve orientation by unloading a piece from a machine rather than having the piece blown out with a blast of air and caught in a bin, the machine cycle may have to be extended and the throughput rate

Fig. 1. Overview of experimental robot system. Six degree-of-freedom (3D, 3R) robot holds the parallel jaw gripper. Two overhead cameras view bins.

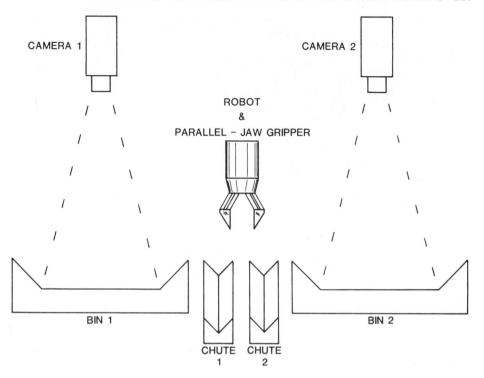

Fig. 2. Schematic of dual-bin system configuration.

reduced. Other mechanisms to unload workpieces while preserving orientation may be expensive and space consuming. Pallets which are used to maintain the orientation of parts are susceptible to being upset during transfer from one work-station to another. Many machines are already set up to accommodate batch production and use stacks of bins for intermediate storage. Parts which come from vendors, long-term storage, or distant warehouses are usually shipped unoriented. The cost of storing or shipping oriented parts, due to low packing density, is usually prohibitive.

Due to the disadvantages of these three alternative approaches, robots with vision will be applied in the future to feeding workpieces. Their use will make it easy to change from one part to another for batch changes.

Various technical contributions have been made to the bin-of-parts problem. A data base of image of castings has been made available[2]. A number of studies have been made which estimate the position and orientation of workpieces on flat surfaces, such as belt conveyors[3]-[10]. The acquisition of a hanging part has been studied[11]. Dihedrally tipped boxes which vibrate have been used to orient isolated parts[12]. Overlapping parts, which are nearly horizontal, have been analysed for position and orientation[13], [14]. Regions which belong to the same object have been identified in an image of a stack of blocks[15], [16]. The tasks of estimating the pose of pieces in the bin which were partially occluded and of acquiring only pieces which could be transported directly to a goal with a prespecified pose have been examined[17]. Ellipses have been used to locate the circular vacuum cleaner filter which is on the 'top' of a pile of similar parts[18]. The parts of an assembly have been separated by a robot from a heap by grasping for protrusions which were located visually and by pushing with the hand at various levels[19]. Heuristic visual methods for acquiring workpieces in bins have been described[20]. Electromagnets have been dragged through bins to acquire billets[21].

Functional experimental robot systems employing special vacuum cup hands and simple vision algorithms have been described[22], [23]. These systems had an integrated architecture which permitted all pieces to be transported to a goalsite with a prespecified orientation, regardless of which flat surface the vacuum cup acquired.

This paper describes part of the continuing evolution of the concepts embodied in these systems. The experimental robot system discussed here uses vision and a parallel jaw gripper to acquire randomly oriented cylindrical workpieces piled in bins. This system represents another step forward in the technology of handling pieces supplied in bins. The generality of the system design is particularly well-suited for batch applications.

System architecture

The architecture of the robot system to perform bin picking and orienting was as follows: a supply bin, and overhead camera, a four degree of freedom robot, a parallel jaw gripper, and a receiving chute. The experimental robot system is shown in Fig. 1; the schematic is given in Fig 2. The receiving chute could be used to deliver oriented pieces mounted, for example, on belt conveyors as depicted in Fig. 3. As the belts are indexed, the pieces might pass through unloading stations where transfer robots transfer them into machine fixtures.

The supply bins used for this system were designed to have a flat bottom area the same size as the field of view of the overhead camera. The sides of the bins were sloped to eliminate the possibility of the parallel jaw gripper structure colliding with the sides of the bin when the gripper attempts to acquire pieces resting on the bottom of the bin near the edge of the field of view. The bins were designed to accommodate piles of pieces having significant depth; hence the concern about collisions.

The overhead cameras were mounted in a position approximately above the centre of each bin. The actual alignment of the cameras relative to the robot was determined by a semi-automatic software procedure, camera calibration. This procedure is performed after each camera has been aimed. The relationship between cameras and the robot is held fixed throughout the experiment.

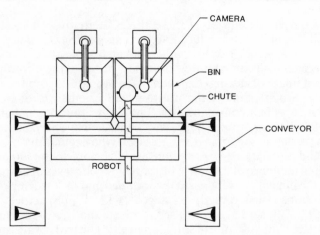

Fig. 3. System floor plan. Possible belt conveyor layout is depicted.

The six degree-of-freedom robot was limited to four degrees of freedom for the experiment; one rotational degree of freedom, rotation about the vertical axis, was needed to align the jaws of the grippers with the axes of the cylinders. Three continuous degrees of freedom in position were required since the depths of the cylinders in the pile were not known *a priori*.

The parallel jaw gripper was symmetric by design and, therefore, for orienting cylinders, could be restricted to rotations no greater than a quarter turn from a reference angle. Because of the parallel grasping surfaces, grasped cylinders have only one unknown orientation angle, the angle of rotation in the plane defined by the parallel surfaces. The receiving chute was an inclined V-groove. Gravity and the design of the chute assured that all cylinders which were placed on the chute by the robot were aligned when they reached the bottom of the chute.

Parallel jaw gripper

The gripper should be rugged, simple, and versatile. That is, it should be able to work in a production environment, not be too complex to affect reliability, and yet be able to handle a family of pieces, at least cylinders and the pieces obtained in the subsequent stages of the manufacturing process.

A vacuum cup gripper like the University of Rhode Island 'surface adapting vacuum gripper' was initially considered[23]. Such a gripper could be equipped with interchangeable elongated vacuum cups. The ability to adapt the vacuum cup to unknown grasping surface angles is necessary for grasping pieces in a bin.

This kind of gripper was rejected for several reasons. The holding force of a vacuum cup may be marginal for high speed movement of pieces since the grasping pressure is limited. The grasping performance of the vacuum cups with different size workpieces was uncertain. Rigidly fixing the workpiece pose relative to the gripper is difficult due to the compliance of the cup. Heat and dirt adversely affect the ability of vacuum cups to form adequate seals.

The remaining alternatives were considered in the context of clamping grippers. One alternative was to use finger tips with V-notches rather than flat pads. The V-notch approach was rejected although it conceivably could reduce workpiece pose (position and orientation) uncertainty in the hand and hold cylindrical work-pieces more stable than flat pads. V-notches were rejected for the following reasons. They were not generally applicable to other kinds of workpieces. V-notches require thick fingers and to acquire closely packed pieces, finger thickness should be minimised. To be effective, the size of the V-notch should be proportional to the size of the workpiece. If the pieces did not always align with the direction of the notch, the complexity of a swivel degree of freedom would be needed to align the notch with the piece.

A single actuator system was selected to control the position of both fingers. The use of independent finger position control was rejected because the class of work-pieces being considered would not be damaged by motion against other workpieces during the process of acquisition. Furthermore, this motion is minimised when an accurate camera calibration is available to locate the piece visually. To obtain programmable finger openings for different size pieces, a position control system using a motor encoder was selected. The use of an air cylinder, for example, was rejected since the finger opening would have to be programmed with mechanical stops. Multibin applications would have been difficult to implement.

The fingertips were designed without the ability to accomplish controlled reorientation of the piece in the gripper. Such a capability was rejected because this would require increasing the overall weight of the gripper and the size of the finger-tips. For bin-picking applications, the fingertips must be small enough to get

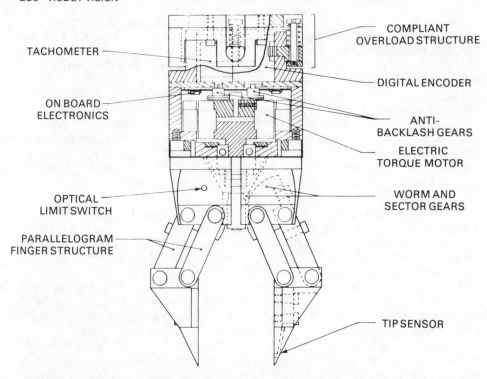

TACHOMETER

ON BOARD
ELECTRONICS

OPTICAL
LIMIT SWITCH

PARALLELOGRAM
FINGER STRUCTURE

COMPLIANT
OVERLOAD STRUCTURE

DIGITAL ENCODER

ANTI-
BACKLASH GEARS

ELECTRIC
TORQUE MOTOR

WORM AND
SECTOR GEARS

TIP SENSOR

Fig. 4. Sectional view of parallel jaw gripper. Component locations are indicated.

between closely packed pieces. The tips should be mechanically smooth with no projections which could get caught on pieces. Further, the tips should be only wide enough to keep large off-centre pieces from rotating so that a minimum clearance space is needed for the fingers to grasp the piece. Finally, because the application did not require certain kinds of information for its successful operation, force or torque sensors on the wrist or on the fingers and fingertip contact sensors were not employed.

Mechanical structure of the parallel jaw gripper
The gripper structure consists of a cylindrical housing and two parallel fingertips. A sectional view of the gripper is shown in Fig. 4. A top-down description keyed to the figure follows.

The gripper is attached to the wrist by means of the top plate of the compliant overload structure. This structure holds the cylindrical housing using four pre-loaded compression springs. Overload forces and torques on the gripper are relieved with automatic resetting when the cause of the overload is removed.

The upper section of the cylindrical housing contains a digital position encoder and a tachometer. A mounting plate also serves to hold the printed circuit board for the sensor electronics. The lower section houses the dc torque motor. Antibacklash gears are used to drive the tachometer and encoder above.

The motor drives the fingers through a worm and sector gears. The parallelogram finger structure is attached as follows: the pairs of outer linkages are free to rotate at both ends; the inner linkages are fixed to the sector gears at the upper end and are free to rotate at the lower end. The linkages are attached to finger support blocks;

the fingertips attach to the support blocks. This feature permits easy finger tip replacement.

Sensors

The following sensors were selected for this application. A tachometer was necessary for the position control system. (An existing position servo controller was used.) Optical limit switches were used to sense the presence of an overload condition. The overload signal was used to alter the execution of the piece pick-up program. An optical limit switch was also employed to establish a reference opening position for the fingers (see Fig. 4). The finger opening was determined by reading the incremental position encoder on the motor and evaluating a function to determine the opening of the fingers. An inverse relationship was used to set the motor position encoder to obtain a particular finger opening. Measurement of finger opening could also be used for in-process inspection or to determine which side of a rectangular bar is being grasped, for example.

The fingertips were equipped with an infrared transmitter–receiver pair which provided a light beam between the two tips (see Fig. 4). To obtain reliable performance, the transmitter was pulsed (10 per cent duty cycle) and the receiver was synchronised with the pulses. The interruption of the beam was used to detect the instant when a piece passed between the fingertips as well as to verify that a piece was present when the fingers were closed.

Gripper software

Gripper software had three functions: position initialisation, fingertip distance setting and measuring, and pick-up sequencing. Position initialisation was required because incremental position encoders were used. Position initialisation established a reference zero position by the following actions. The fingers were opened until a limit switch closure was sensed. The encoders were zeroed and then commanded to close by a fixed offset distance. The encoders were zeroed once again to establish the reference zero relative to the limit switch position.

Fingertip distance setting and measuring were accomplished indirectly since the position encoder was mounted on the motor shaft and not on the sector gears. It would have been possible to compute the functional relationship between motor position encoder counts and fingertip distance. However, an alternative procedure was employed which used calibrated blocks to obtain data points of distance versus encoder counts from which polynomial approximations were generated. A polynomial of degree four was used to determine the number of encoder counts necessary to obtain a given fingertip opening. A polynomial of degree three was used to determine the actual fingertip opening given the encoder count reading. Employing the least square error polynomial approach eliminated the need to verify the mechanical dimensions and alignments which are assumed in any geometric analysis. Further, any deviations from the ideal were automatically accounted for by the curve fitting approach whereas a new geometric analysis would have been required.

The gripper pick-up sequence began by obtaining a schedule of three fingertip distances, the line-of-sight vector space coordinates, and the direction of the gripper opening. The rotated fingers were opened to distance one. The fingertips were moved down along the line-of-sight vector toward the holdsite until the fingertip sensor was interrupted. The fingers were then opened to distance two and the downward motion was continued an additional grasping distance. The fingers were closed to distance three and then the gripper was moved up to the traverse plane. This simple procedure accomplished the acquisition of pieces having unknown depths in the bin.

Vision analysis

Vision analysis of the bin image was divided into two parts. The first part checked the image for the presence of objects which needed to be further analysed. This check was performed to determine the absence of workpieces. The second part continued the analysis to select a holdsite location and direction. This analysis is performed within the task PICTUR.

Vision analysis was initiated when a grey scale image of the pieces in the bin was obtained. The intensity value of each pixel (picture element) was compared with a common threshold to create a binary image. The first operation in the analysis was the application of a shrinking operator. The shrinking operator was applied to the successive images to reduce the total number of pixels which were collected for holdsite selection. These pixels were clustered and sorted by size. The largest cluster was selected and the centroid (centre of gravity), eigenvectors (axes of minimum and maximum moments of inertia, through the centre of gravity) and eigenvalues (moments of inertia about these axes) of that cluster were computed. The location and direction of the gripper fingertip opening relative to the image was obtained from the centroid location and the major eigenvector direction (axis of minimum moment of inertia). The ratio of eigenvalues determined which fingertip opening schedule was used: cylinder on end, or cylinder inclined or flat.

A vector in space was defined by the image location of the centroids of the selected cluster. The vector is the projection of the centroid along the line-of-sight. During acquisition, the fingertips of the gripper travelled along the line-of-sight path.

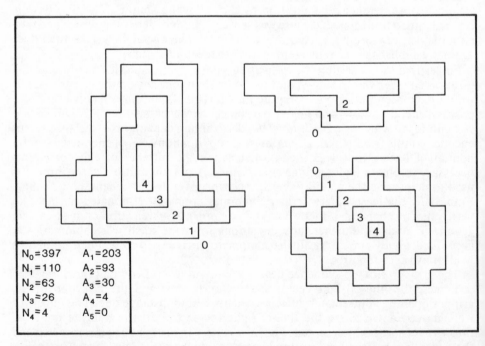

Fig. 5. Shrinking procedure example. Pixels are labelled with their ultimate values. Intermediate totals are listed in table.

The shrinking procedure was performed as follows: given a grey scale image G and a threshold T, create a two-valued image S_0. Specifically,

$$S_0(i,j) = \begin{cases} 1, & G(i,j) \geqslant T \\ 0, & \text{otherwise.} \end{cases}$$

S_0 has a shrink level of zero. Define an indicator function δ_k $(k=1,2,\ldots)$ as

$$\delta_k(i,j) = \begin{cases} 1, & S_{k-1}(i,j) = k \\ 0, & \text{otherwise.} \end{cases}$$

Given k-valued image S_{k-1}, the $(k+1)$-valued image S_k is created by incrementing all pixels having the value k which are totally surrounded (in the eight-neighbour sense) by pixels having the value k. That is,

$$S_k(i,j) = \begin{cases} k+1, & \sum\limits_{l=-1}^{1} \sum\limits_{m=-1}^{1} \delta_k(i+l, j+m) = 9 \\ S_{k-1}(i,j), & \text{otherwise.} \end{cases}$$

Let N_0 be the number of zeros in the image and A_1 be the number of ones. Then

$$N_0 = \sum_{i,j} [1 - \delta_1(i,j)]$$

and

$$A_1 = \sum_{i,j} \delta_l(i,j)$$

The total number of $(k+1)$ in the image S_k is

$$A_{k+1} = \sum_{i,j} \delta_{k+1}(i,j)$$

and the total number of k is

$$N_k = A_k - A_{k+1}$$

An example of the application of shrinking to a 20×30 pixel image is shown in Fig. 5. The progression shown in the A_k list demonstrates the effect of shrinking on the number of active pixels. After shrinking one level, the number of active pixels is 93, a reduction of 110 from the original 203. The 93 pixels include all of those labelled 2, 3 or 4. Notice that after two levels of shrinking, only two active clusters remain.

Shrinking is terminated when the image S_L has the properties that

$$N_L \leqslant \text{MAXPIX} < N_{L+1}$$

and

$$0 < \text{MINLEV} \leqslant L \leqslant \text{MAXLEV},$$

where MAXPIX is a parameter which controls the number of pixels given to the clustering program and MINLEV, MAXLEV are parameters which bound the shrinking effort.

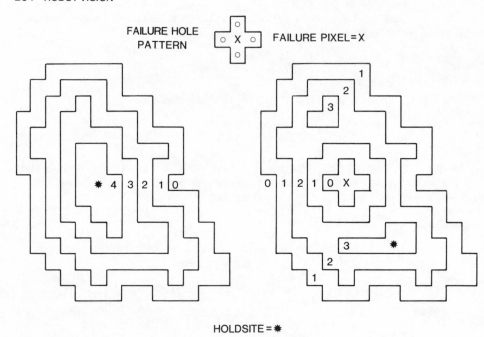

HOLDSITE = ✻

Fig. 6. Effect of failure hole. Before shrinking, a failure hole pattern was put into the right hand figure at the location corresponding to the holdsite marked on the left figure.

Clustering time grows with the square of the number of active pixels. If N_{MAXLEV} > MAXPIX, this means that the field of view is too large and should be reduced. Such would be the case for small pieces when too many pieces fill the field of view. A way to deal with the latter problem is to restrict the number of active pixels to those within a sub-image window. The maximum shrinking level is chosen such that all good pieces are gone and only foreign objects remain.

When shrinking was terminated, the N_L pixels having value L were clustered by distance. The clustering started with N_L clusters. Two pixels were put in the same cluster if the (city-block) distance between them was less than the clustering distance. The resulting clusters were sorted according to size. If the sizes were above a minimum MINPIX, holdsites were computed for the three largest clusters; the primary holdsite for the parallel jaw gripper. The ratio of the eigenvalues was used to distinguish between cylinders which were on end (ratio almost one) and those which were not. The appropriate fingertip opening distance schedule was then used.

In the event that the largest cluster was smaller than MINPIX, a window was placed about its centre of gravity location and the clustering process was repeated using only pixels having value $L - 1$. If this procedure did not result in any clusters larger than the minimum size MINPIX, no holdsite was selected. The selection failure was noted and the vision analysis was terminated. When selection failures occurred in all windows in the cycle a message to the operator was generated.

The locations of holdsites where unsuccessful acquisition attempts occurred were kept in a short-term memory. To prevent repeated unsuccessful attempts to acquire the same piece in the same way, a failure hole pattern was put into the binary image S_0. The pattern was centred on the failure location pixel and consisted of replacing the centre pixel and its four-neighbours with zeros. The effect of the failure hole was to cause subsequent holdsites to be selected some minimum distance from the unsuccessful one. An example of the shifting of a hold-site when a failure hole is put into a pixel cluster is shown in Fig. 6. Notice that in this case, the failure hole pattern caused two clusters to be created which were approximately equidistant from the failure holdsite and the periphery of the original cluster.

Robot vision in general requires knowledge of spatial projections of image points. Given a known plane, the coordinates of the piercing point of a line-of-sight projection vector is easily determined. Then spatial quantities such as distance, angle, and area can be computed from image data.

Relating image measurements to spatial measurements requires an accurate camera model. When angular measures (light ray to light ray) are interpreted as linear ones, spherical distortion errors result, even if the lens is perfect. In practical applications, it is desirable to use the entire field of view of the camera out to the edge and not be restricted to the central zone and give up resolution. For these experiments, a least-square-error second-order polynomial approximation was used.

The line-of-sight vector was obtained by computing the coordinates of two piercing points in planes fixed relative to the camera. The geometry of the calcula-

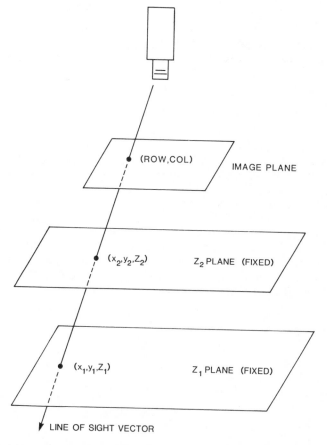

Fig. 7. Line-of-sight vector geometry. Image plane points and projected through fixed calibration planes to define a space vector.

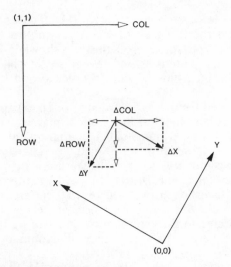

Fig. 8. Camera calibration. Geometry of the image plane and calibration plane coordinate systems is shown. Sampling grid and approximate interpolation formula is given.

tion is shown in Fig. 7. Given the row and column location of an image point, the coordinates of the piercing points in the fixed planes Z_1 and Z_2 were computed. The values of x_1, y_1, x_2, and y_2 were computed by evaluating individual functions of the form

$$V = aR^2 + bC^2 + cRC + dR + eC + f$$

where V is the coordinate value x_1, y_1, x_2, or y_2, R is the row coordinate, C is the column coordinate, and (a, b, c, d, e, f) are the least-square coefficients.

The least-square coefficients were obtained by collecting sample points in both planes Z_1 and Z_2. A regular $n \times n$ sampling grid was used. Because the robot coordinate axes (X and Y) might not be aligned with the camera coordinate axes (ROW and COL), test moves were made with a circular calibration chip at the centre of the field of view. These moves provided an interpolation formula which allowed the calibration chip to be placed roughly at a particular sample point. The interpolation formula was obtained by moving the chip in the X-direction only and measuring the motion of the chip image, then repeating the test for the Y-direction as shown in Fig. 8. These motions are related as

$$\Delta R = S_{11} \Delta X + S_{12} \Delta Y$$
$$\Delta C = S_{21} \Delta X + S_{22} \Delta Y$$

where

$$S_{11} = (\Delta R / \Delta X)_{\Delta Y = 0}, \; S_{21} = (\Delta C / \Delta X)_{\Delta Y = 0},$$
$$S_{12} = (\Delta R / \Delta Y)_{\Delta X = 0} \text{ and } S_{22} = (\Delta C / \Delta Y)_{\Delta Y = 0}$$

By inverting this relationship, an interpolation formula was obtained which yielded an approximately uniform image plane sampling grid when the chip was moved according to

$$\Delta X = (S_{12}\Delta C - S_{22}\Delta R)/D$$
$$\Delta Y = (S_{21}\Delta R - S_{11}\Delta C)/D$$

where $D = S_{12}S_{21} - S_{11}S_{22}$.

Once the n^2 samples in each plane had been taken, the data were used to obtain the least-square-error set of coefficients for each coordinate in each plane. A formal presentation of this computation follows. Denote the row value by R, the column value by C, and the corresponding coordinate value x_1, y_1, x_2, or y_2 by X. Define the image row-column data vector as

$$\mathbf{J}_i = (R^2, C^2, RC, R, C, 1), i = 1, \ldots, n^2.$$

Define the coefficient vector for the corresponding coordinates x_1, y_1, x_2, or y_2 as

$$\mathbf{A} = (a, b, c, d, e, f)^T.$$

For each coordinate data point

$$X_i = \mathbf{J}_i\mathbf{A}, i = 1, \ldots, n^2$$

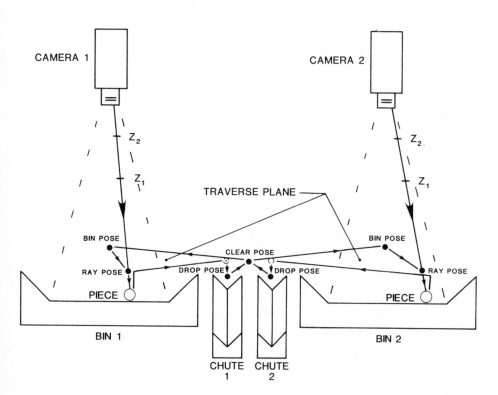

Fig. 9. Schematic of dual-bin execution cycle. Key poses are labelled.

Then for all the data points

$$X = [J]A$$

where $X = (n \times 1)$-vector of X_i and $[J] = (n \times 6)$-matrix with rows J_i. The least-square coefficient vector A is given by

$$A = ([J]^T[J])^{-1} [J]^T X$$

where the matrix product gives the usual pseudoinverse matrix.

Top level software

Experiments were performed using both a single bin of pieces and two bins with different pieces. The software description given here is for the dual bin configuration, the single bin is similar. The execution cycle for this case is illustrated in Fig. 9. The cycle starts when the gripper is at the CLEAR pose above the chute and a 'clear' signal is issued. An image of the pieces in a bin, say bin 1, is brought into the computer and the first part of the image analysis is begun. Assuming the previous image analysis for the other bin, bin 2, was completed, a 'ready' signal causes the gripper to be sent directly to the RAY pose in the TRAVERSE plane. If the 'ready' signal has not yet been issued, to minimise motion delays the gripper is sent toward the centre of the bin, BIN pose, while the analysis continues. (The gripper is redirected to the RAY pose as soon as the 'ready' signal is received). The pick-up sequence is initiated starting at the RAY pose. The gripper is aligned with the hold-site direction and is sent down along the line-of-sight path. Piece pick-up is performed and the gripper moves up to the TRAVERSE plane. Then the gripper moves to a pose above the DROP pose and verifies that a piece is in the gripper. If a piece is held, the gripper moves to the DROP pose close to the chute, drops the piece, and then moves to the CLEAR pose. If no piece is held, the gripper moves directly to the CLEAR pose. A 'clear' signal is issued and the second half of the cycle is executed with an image of the bin just picked being brought into the computer for analysis and the gripper being sent to the bin previously analysed. The nominal execution cycle continues this pattern of alternating analysis followed by pick-up in each bin.

The software was organised as four concurrent tasks which ran under the control of a multitask real-time execution system. The first of the four tasks is the main task, MTASK, which initiates the dual bin picking cycle. The cycle described in the preceding paragraph is actually composed of three tasks: NEAR, PICTUR, and DPICK.

The flowchart for the main task MTASK is shown in Fig. 10. First the arm is initialised by establishing reference zero positions for all degrees of freedom including the fingertip opening. Next, the arm is commanded to move the gripper to the CLEAR pose. The task NEAR (CLEAR) is invoked and execution of MTASK is suspended until a 'near' signal is received. Parameters are set to cause the cycle to start by analysing the bin 1 image and to cause the gripper to wait for the analysis to be completed by signalling that image 2 was bad. MTASK then starts the tasks PICTUR and DPICK with DPICK being assigned a higher priority. Then MTASK terminates.

The task NEAR flowchart is given in Fig. 11. This task checks how near the pose of the robot is to a specified destination pose and issues a signal when the pose is

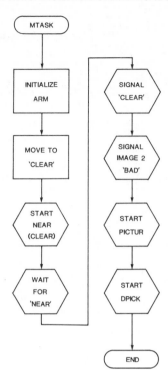

Fig. 10. Task flowchart: MTASK.

'near'. The task waits a few ticks (10ms) of the real-time clock before reading the arm joint coordinates and computing the distances from the arm joint destinations. These distances are compared against a list of tolerances for the desired pose. If any distance is too large, the procedure is repeated. When all of the distances are within the tolerances, a 'near' signal is issued and the task is terminated.

The flowchart for task PICTUR is shown in Fig. 12. This task performs all the image analysis and supplies the line-of-sight and holdsite data to the task DPICK. The task begins by getting the number of the bin to be analysed. When a 'clear' signal is received, the image of the bin is brought into the computer and the first part of the image analysis is performed. A test is made for image quality and an image quality signal, good or bad, is issued. If the quality is bad, the task jumps to set up the next bin image analysis and restarts the task. If the image quality is good, the second part of the image analysis is performed. The line-of-sight-path and holdsite data are computed and then a 'ready' signal is issued. Finally, the next bin image analysis is set up and the task restarted.

The flowchart for task DPICK is shown in Fig. 13. This task performs all of the robot motions; in particular, it executes the piece pick-up sequence. The task begins by getting the number of the bin to be picked and checks the quality of the image. If the image quality is bad, the arm is commanded to move to the CLEAR pose in preparation to taking another picture. If the image quality is good, the arm is commanded to move to the BIN pose. When a 'ready' signal is received, the arm is commanded to the RAY pose in the TRAVERSE plane. The 'pick-up piece' operation is performed by sending the gripper down along the line-of-sight path. The gripper is then moved up to the TRAVERSE plane, and commanded to move

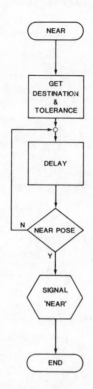

Fig. 11. Task flowchart: NEAR.

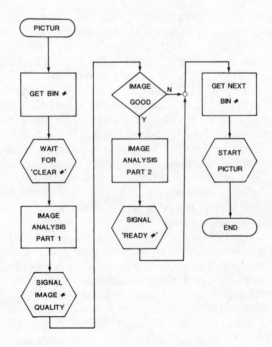

Fig. 12. Task flowchart: PICTUR.

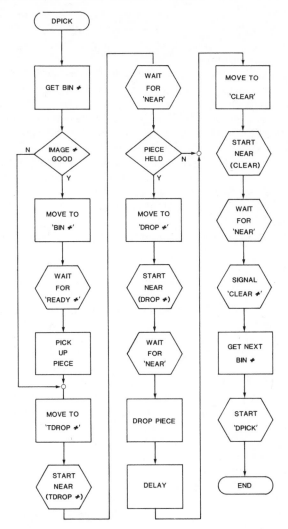

Fig. 13. Task flowchart: DPICK.

in the plane over the DROP pose. When a 'near' signal is received, the presence of a piece in the gripper is tested. If there is no piece, the arm moves directly to the CLEAR pose. If a piece is held, the arm is moved to the DROP pose, the piece is dropped into the chute and then is commanded to move to the CLEAR pose. When a 'near' signal is received, a 'clear' signal is issued, the next pick-up task is set up and the task restarted.

Experimental results

Experiments were performed using the robot system shown in Fig. 1. The system employed solid-state cameras and a parallel jaw gripper. Only four degrees of freedom were needed by the arm: three for position and one to rotate about the vertical gripper axis. The system architecture was that shown in Fig. 2 with experiments performed using both a single bin of cylinders and two bins with a different size in each.

The robot arm moved slowly in comparison to current industrial robot standards. Linear axis motion did not exceed 30cm/s and rotary motion did not exceed 90°/s. Titanium cylinders 6cm × 1.5cm diameter and 7.6mm × 3cm diameter were used in the experiments.

Cycle times to acquire a cylinder and deliver it to a receiving chute ranged from 8 to 10s when a single supply of one size was used. The major cause of cycle time variation for the single-bin system was the delay resulting from the image analysis time required to select a holdsite. This source of delay was substantially masked in the dual-bin system because more arm motions were overlapped with the image analysis task. By using a dual supply configuration with one bin of each size, the cycle times for one cylinder from alternate bins ranged from 5.5 to 7.5s per piece. Other sources of cycle time variations came from variations in the distance the arm travelled due to the location of the selected holdsite in the bin and variations in the distance travelled along the line-of-sight path due to the actual depth in the pile of the piece. Acquisition failures also added to the overall cycle time. However, the statistical effect of these failures was small in the experiments run.

Conclusion

The feasibility of robots employing vision to acquire randomly oriented cylinders has been demonstrated for the first time. An experimental robot system using vision and a parallel jaw gripper was able to acquire randomly oriented cylindrical workpieces piled in bins and deliver them to a V-chute which discharged oriented cylinders. Binary image analysis was adequate to guide the gripper into the multi-layered piles. Complementary information was provided by simple sensors on the gripper.

In addition to using a faster arm, cycle time could have been improved by using the centres of the analysis windows for the location of the BIN poses instead of the bin centre. This might be particularly important for single-bin application. Also, cycle time could have been decreased by reducing the distances over which the arm was moved slowly such as during the pick-up procedure.

This system represents an initial step toward giving robots the capabilities needed for future automation applications, especially in small batch production. The general bin picking problem is still to be solved; however, the system described demonstrates the effectiveness of combining heuristic vision algorithms and sensor equipped grippers to handle an industrially important class of workpieces and to perform the task 'on time.'

Substantial work remains to be done in applying the insights derived from this experience to other applications; results for other types of workpieces will be reported. Many questions have been framed as a result of this. For example, what is the role of compliance in the gripper structure relative to compensating for short-comings in the image analysis or the vision system? What sensors should be mounted on the gripper to enable the same robot to perform a variety of tasks without changing grippers? This could be particularly important if it is desired to perform these tasks within the same production cycle.

The analysis task for an isolated piece in the robot hand is much simpler than when that piece is seen with other similar pieces in a bin. For this reason, the major goal of the bin image analysis is to extract enough information to allow some piece to be acquired successfully. Many pieces still possess too much pose uncertainty when acquired to permit the piece to be placed at a goalsite, dropped onto a surface, or fed to a chute. This additional pose information can be obtained by analysing the image of the piece held by the robot. The drawback of this approach is that the need for a second look at the piece usually requires a second camera system or arm motions which are performed solely to provide an unclutttered view of the piece held in the hand. An area of future work is to determine ways to avoid the need for the second camera system and the associated arm motions.

Acknowledgement

The robot system described here came to exist through the efforts of the Robot Research Group, past and present. Special recognition is given to James Hall who kept the system healthy. The constructive comments by the reviewers which helped to improve the paper are gratefully acknowledged.

References

[1] A. Pugh, K. Waddon and W. Heginbotham, 'Orientation and inspection of component parts,' SPIE vol. 145. Industrial Applications of Solid-State Image Scanners, London, UK, pp. 66–67 (Mar. 1978).

[2] M. Baird, 'A computer vision data base for the industrial bin of parts problem,' *General Motors Res. Publication*, GMR-2502 (Aug. 1977).

[3] M. Baird, 'Image segmentation technique for locating automotive parts on belt conveyors,' *Fifth Int. Joint Conf. Artificial Intelligence*, Cambridge, MA, pp. 694–695, Aug. 22–25, 1977.

[4] W. Heginbotham, D. W. Gatehouse, A. Pugh, P. W. Kitchen and C. Page, 'The Nottingham "SIRCH" assembly robot,' *Proc. First Conf. Industrial Robot Tech.*, Nottingham, UK, pp. 129–142 (Mar. 27–29, 1973).

[5] P. Martini and G. Nehr, 'Recognition of angular orientation of objects with the help of optical sensors,' *The Industrial Robot*, vol. 6, pp. 62–69 (June 1979).

[6] R. Nagel, G. VanderBrug, J. Albus and E. Lowenfeld, 'Experiments in part acquisition using robot vision,' National Bureau of Standards, CMEPT, Mechanical Processes Div., Automation Tech. Sec., Washington, DC (Oct. 1979).

[7] C. Rosen *et al.*, 'Exploratory research in advanced automation,' Fourth Rep., Stanford Res. Instit., Menlo Park, CA (June 1975).

[8] C. Rosen *et al.*, 'Machine intelligence research applied to industrial automation,' Sixth Rep., Stanford Res. Instit., Menlo Park, CA (Nov. 1976).

[9] Y. Tsuboi, T. Shiraishi and N. Kosaka, 'Positioning and shape detection algorithms for an industrial robot,' *Syst. Comput. Controls*, vol. 4, pp. 8–16 (1973).

[10] G. VanderBrug, J. Albus and E. Barkmeyer, 'A vision system for real time control of robots,' *Proc. Ninth Int. Symp. on Industrial Robots*, Washington, DC, pp. 213–231, Mar. 13–15, 1979.

[11] Y. Tsuboi and T. Inoue, 'Robot assembly system using TV camera,' *Proc. Sixth Int. Symp. on Industrial Robots*, Nottingham, UK, pp. BB: 21–32, Mar. 24–26, 1976.

[12] D. Grossman and M. Blasgen, 'Orienting mechanical parts by computer-controlled manipulator,' *IEEE Trans. Syst., Man, Cybern.*, vol. SMC-5, pp. 561–565 (Sept. 1975).

[13] J. Dessimoz, M. Kant, J. Zurcher and G. Granlund, 'Recognition and handling of overlapping industrial parts,' *Proc. Ninth Int. Symp. on Industrial Robots*, Washington, DC, pp. 357–366, Mar. 13–15, 1979.

[14] W. Perkins, 'A model-based vision system for industrial parts,' *IEEE Trans. Comput.*, vol. C-27, pp. 126–143 (Feb. 1978).

[15] G. Falk, 'Interpretation of imperfect data as a three dimensional scene,' *Artificial Intelligence*, vol. 3, pp. 101–144 (1972).

[16] A. Guzman, 'Decomposition of a visual scene into three dimensional bodies,' *Proc. AFIPS Fall Joint Comput. Conf.*, vol. 33, pp. 291–304 (Dec. 1968).

[17] R. Kelley, J. Birk and L. Wilson, 'Algorithms to visually acquire workpieces,' *Proc. Seventh Int. Symp. on Industrial Robots*, Tokyo, Japan, pp. 497–506, Oct. 19-21, 1977.

[18] S. Kashioka, S. Takeda, Y. Shima, T. Uno and T. Mamoda, 'An approach to the integrated intelligent robot with multiple sensory feedback: visual recognition techniques,' *Proc. Seventh Int. Symp. on Industrial Robots*, Tokyo, Japan, pp. 531–538, Oct. 19–21, 1977.

[19] A. Ambler, H. Barrow, C. Brown, R. Burstall and R. Popplestone, 'A versatile computer-controlled assembly system,' *Third Int. Joint Conf. on Artificial Intelligence*, Stanford, CA, pp. 298–301, Aug. 20–23, 1973.

[20] J. Birk, R. Kelley and L. Wilson, 'Acquiring workpieces: three approaches using vision,' *Proc. Eighth Int. Symp. on Industrial Robots*, Stuttgart, West Germany, pp. 724–733, May 30–June 1, 1978.

[21] A. Ferloni, I. Franchetti, P. Vicentini and P. Fici, 'ORDINATORE: a dedicated robot that orientates objects in a predetermined direction,' *Proc. Tenth Int. Symp. on Industrial Robots*, Milan, Italy, pp. 655–658, Mar. 5–7, 1980.

[22] R. Kelley, J. Birk, D. Duncan, H. Martins and R. Tella, 'A robot system which feeds workpieces from bins into machines,' *Proc. Ninth Int. Symp. on Industrial Robots*, Washington, DC, pp. 339–355, Mar. 13–15, 1979.

[23] J. Birk, R. Kelley and H. Martins, 'An orienting robot for feeding workpieces stored in bins,' *IEEE Trans. Syst., Man. Cybern.*, vol. SMC-11, pp. 151–160 (Feb. 1981).

FLEXIBLE ASSEMBLY MODULE WITH VISION CONTROLLED PLACEMENT DEVICE

W. B. Heginbotham, D. F. Barnes, D. R. Purdue and D. J. Law, PERA*, UK

Presented at the 11th International Symposium on Industrial Robots, 7-9 October 1981, Tokyo, Japan.
Reprinted by kind permission of the Japan Industrial Robot Association.

Abstract

The paper describes a prototype system for the feeding, inspection and assembly of small parts. The machine can be adjusted so as to feed different components by way of a simple programming method. This versatility, combined with the ability to reject scrap components before they reach the assembly station, can be a valuable aid to automated small batch assembly, magazining for in-process handling, sorting and inspection.

Introduction

Two important aspects of automatic assembly are, firstly, the on-line inspection of every component to ensure that the assembly machine is not damaged by rogue components, and the quality of the assembly itself is not lost; secondly, the flexibility of the feeding equipment to allow for small batch production and rapid re-tooling between batches.

A more controversial aspect of automation in general is how can machine vision be best applied to the assembly process; how sophisticated should that vision be and at what point in the process should it be used.

This paper describes a system consisting of bowl feeder, machine vision and pick-and-place robot, that is an extension of earlier work,[1, 2] which investigated the use of opto-electronics to sense components as they escaped from a vibratory bowl feeder. A microprocessor (LSI11) was used to decide from simple binary visual data:

○ Was the component the correct one?
○ Were all the essential features of the component present, i.e. holes, steps, etc.
○ What was the orientation of the component in relation to two mutually perpendicular reference planes formed by the floor and wall of the feeder flight?

There was a minimum requirement for mechanical bowl tooling, the major part of the 'tooling' being performed by an operator using an interactive VDU. By using

*PERA is the trademark of the Production Engineering Research Association of Great Britain.

such a display, no special bowl tooling expertise was necessary but some judgement was required by the setter in order to determine suitable interrogation points.

The intention of this further development is to stimulate a real assembly situation, by the addition of a pick-and-place device, and show how machine vision can be effectively applied to the industrial problem of automated small batch assembly, as well as being useful in the more general field of inspection and sorting.

The vision system is relatively simple compared with some current experimental work elsewhere, and some commercially available systems. However, this simplicity has the advantages of fast recognition times, and a simple method of programming, together with a lower overall cost. Machine vision need not be sophisticated, if it is designed for a particular area of application, and a concerted effort is made to optimise lighting, camera arrangement and, in this case, the component transport mechanism.

Description of the operation of the assembly module

The component used to demonstrate the operation of the assembly module is a bicycle brake lug, which is illustrated in Fig. 5.

Referring to Fig. 1, parts leave the bowl feeder (A) via the primary control station (B). A single file of components achieved with some simple tooling in the bowl is advanced one component at a time, the components escape onto the belt (C), being detected by the photocell (D) and light-beam from the lamp (E). During their passage up the track of the bowl, the components settle down into a finite number of stable resting states, and these are maintained as each one is processed through the system. Upon reaching the end of the belt, the pusher (F) descends and pushes the part forward and past the inspection station (G). Light from lamp (H) (Fig. 2 and 3) is structured by the shades (I) and falls upon the base of the track, and is reflected

Fig. 1. View of primary control station, belt escapement and pusher mechanism.

Fig. 2. View of placement device and turnaround mechanism.

Fig. 3. Arrangement of selection chutes and VDU.

sideways by the prism (J) onto the side wall of the track. Two thin lines of optical fibres are embedded in the track, one across each wall. These convey a single line plan and elevation view of the component to a semi-conductor linescan camera in the cabinet (V). The constant speed of the pusher allows a 2D silhouette image to be built up from successive scans. Referring to Fig. 5, the bicycle brake lug yields quite an interesting set of silhouettes that uniquely define the different orientations.

Referring to Fig. 2, the next push of the pusher (which has two paddles) moves the part to the pick-up station (K). It is here that a number of different things can happen to the part, according to its orientation. The pick-and-place robot (INA Automation Robomat) is equipped with a 180° turnaround device that rotates about a vertical axis. Referring to Fig. 5, orientations 1 and 2 are interchangeable using the turnaround device, and so are 3 and 4, but 1 and 2 cannot be manipulated into 3 and 4 (and vice versa) because of the limitations of the robot. Only a more advanced robot could achieve that transformation.

Providing that the part is not scrap, and is in one of the two acceptable orientations, then the pick-and-place robot (L) descends and grasps the part with the grippers (M), then ascends and moves across to the template (N) (see also Fig. 3). The template is cut out so as to just allow a part in orientation 1 through it, and represents an assembly station. If the part is 180° out of alignment, i.e. orientation 2, then the turnaround device (O) rotates the grippers as the robot does its horizontal traverse. If the part is scrap, or not in one of the two acceptable orientations, the pick-up station (K) hinges down and the part falls into the chute (P). If it is a scrap component, the trap door (Q) will have already opened and, referring to Fig. 3, the part will be re-routed into the chute (R) that leads to the reject bin (S). If the component is not scrap, but is in an orientation that could not be handled by the robot, i.e. orientations 3 and 4, then it continues down the chute (P) into the 'return bin (T), which collects components that can be sent back to the bowl. Good components, successfully 'assembled', collect beneath the template in another bin (U).

The complete operation of the system is controlled by a DEC LSI11 micro

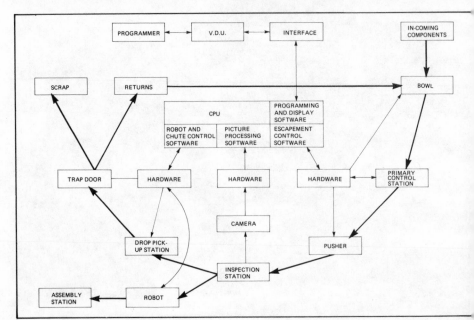

Fig. 4. System block diagram.

computer in the cabinet (V), and programming the system so as to feed a particular part is done via the TV screen (W) and a light pen.

Description of the system architecture

Referring to Fig. 4, thick lines show the flow of components through the system. In-coming parts are loaded into the bowl and escape one at a time from the primary control station onto the belt. When a component reaches the end of the belt, the pusher comes up behind it and moves it past the inspection station at constant speed. A picture of the component is built up from rapid line scans by the camera. The image is then tested by the microcomputer to determine the orientation of the component, or whether it is scrap. According to the result, the component is routed to one of three possible bins.

Referring to Fig. 5, orientations 1 and 2 are handled by the pick-and-place device, orientations 3 and 4 are sent to the 'returns' bin, and would normally be re-circulated automatically in a real situation. Scrap components are routed to the scrap bin.

The thin lines in the figure represent data transfer. The central processing unit carries out four distinct tasks, that of controlling the escapement of parts to the inspection station, the processing of the video data from the camera, and the control of the pick-and-place robot and the two chutes. The fourth task is separate from the other three, being the programming stage, which is done before the system is used in a continuously running situation.

The system software is written using a Macro extension of the LSI11 assembler language (references 1 and 2) which is particularly suitable for muti-task applications and contains some of the commands available in high level languages, whilst still retaining the speed of assembly language programming.

Description of programming method

The system can be taught to recognise the various orientations of a particular component by a simple programming technique, using a light pen and VDU. The program can then be stored for future use. The procedure is to send a part through in a particular orientation, and display the resulting black and white picture on the VDU. Inspection points are then placed onto the picture with the light pen, and the system records whether they lie on a black or white area. The points should be placed so as to characterise areas of interest, in particular the boundary of the object and any internal holes. One of these 'templates' of points should be created for each of the possible orientations.

When the system is in the continuous mode, each picture that is taken is comparted with each of the orientation 'templates' to see if one of them fits, i.e. the inspection points rest on the same shades (black or white) as they did when they were first programmed. If a whole 'template' of points fits, then the orientation is recognised and the component is routed to the appropriate device, i.e. robot or the return bin. If none of the 'templates' fits, then the component is unrecognised and the component scrapped. Tolerance of the component dimensions can be adjusted by how close the points are positioned to the black/white boundaries. Points positioned in the middle of white or black areas guard against the presence of additional holes in the component, or excess material included with the component. Usually, about ten points are sufficient to characterise each orientation.

It should be emphasised that the inspection is a rough one, and cannot detect minor variations in the components, rather, its purpose is to detect things such as the presence or absence of a hole, or a step in diameter; the accuracy of measurements done on the part is of the order of 5% of its size, although it would be possible to improve this.

Once programmed, it is possible to store the set of 'templates' for a particular part and re-load it in the future. When the program has been entered, the system can be run automatically in a stand-alone mode, without the VDU.

System performance

The current prototype system can handle parts at the rate of one every five seconds, this rate being limited by the speed of the robot, and the component transfer mechanism from the bowl feeder to the inspection station. The actual visual processing takes about one-tenth of a second, so that there is a potential for much more rapid operation, if the transfer mechanism were redesigned.

The system is very effective at removing rogue components and miscellaneous scrap. Occasionally, a good component is rejected as scrap, due to the limitations of the transfer mechanism and camera apparatus; however, the possibility of the opposite situation, that is rogue components being placed through the template, can be virtually eliminated by straightforward programming using the inspection point insertion system.

The quality of the picture is, as mentioned above, insufficient for accurate inspection, but sufficient to determine the orientation of a component, and detect coarse irregularities.

Conclusion

A prototype system for feeding, inspection and assembly of small parts has been described. Other applications can be envisaged, such as general inspection and sorting and, in particular, magazining of components prior to loading into an assembly machine.

Two important principles are illustrated by the machine. Firstly, that of a general purpose feeder with a minimum of mechanical tooling, thus making the machine versatile and resistant to mechanical jamming. Secondly, the coarse inspection of all components as they go into an assembly so as to avoid jamming or foul

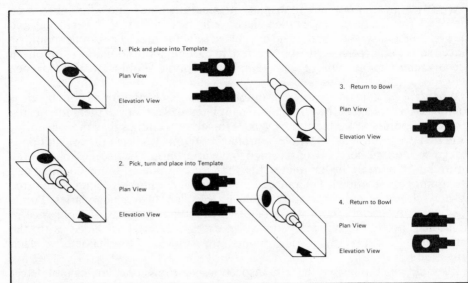

Fig. 5. The four different stable positions of the bicycle brake adjuster.

make-ups in the assembly itself. Occasionally, it happens that a good part is rejected as scrap but this is relatively unimportant. It would be of much more consequence if a bad part or piece of scrap material were inserted into an assembly. During the trials of the prototype at PERA, and whilst being exhibited at the Automan '81 exhibition, the machine never attempted to place scrap into the template, providing it was programmed correctly.

Some limitations are apparent in the prototype machine. Firstly, the escapement of parts from the bowl feeder is too slow, exceeding the actual visual processing by at factor by ten; by re-engineering the transport mechanisms, much faster running could be achieved. Secondly, the accuracy of the inspection may not be high enough for some applications; here again, considerable improvements could be realised in a revised version. Thirdly, the pick-and-place robot is not able to handle some of the component orientations, although by paying attention to what degrees of freedom it is given, more than 50% of components leaving the bowl can be handled, usually more. However, in some applications, it would be desirable to interface a more intelligent and versatile robot, so as to cope with a more difficult range of components.

The three areas of improvement above are well within the bounds of available technology and would be expected to be found on a production version.

Acknowledgements

Grateful thanks are due to: Aylesbury Automation for the loan of a bowl feeder and INA Automation for the loan of a pick-and-place unit. Also to A. Burdett and J. E. Richards for the work they contributed.

References

[1] A. J. Cronshaw, W. B. Heginbotham, A. Pugh, 'Software Techniques for an Optically Tooled Bowl Feeder', *Trans. IEE 3rd International Conference on Control of Automation'*, London (March 1979).

[2] W. B. Heginbotham, A. Cronshaw, A. Pugh, *Proc. 1st Int. Conf. on Assembly Automation,* Brighton, UK, pp. 265-275 (March 1980).

Chapter 6
APPLICATIONS

Applications of robot vision systems by industry are few and far between. Three of the papers describe industrial case studies and the other two are concerned with development aimed at solving specific industrial problems.

VISION SYSTEM SORTS CASTINGS AT GENERAL MOTORS CANADA

Richard D. Baumann, General Motors of Canada Ltd. and
David A. Wilmshurst, RMT Engineering Ltd., Canada (formerly of GM of Canada)

Reprinted from Sensor Review, Vol. 2, No. 3 (July 1982), also published in Engineering Digest, Vol. 28, No. 6 (June 1982).

Abstract

The first production implementation of the General Motors 'Consight' vision system at the St. Catherines, Ontario, foundry is successfully sorting up to six different castings at up to 1,400 an hour from a belt conveyor using three industrial robots in a harsh manufacturing environment (Fig. 2).

Background

The availability of reliable and intelligent industrial robots developed in the mid 1970s is now making itself felt as North American industry adopts them in an effort to improve its competitive position in world markets. The automotive industry has been a fervent advocate of these machines because of the repetitive nature of assembly line work and its inherent monotony.

Until recently robots have required that items to be handled by them should be located accurately and repeatably, since there was no reliable, quick means of informing the robot control system of a change in the location or orientation of a target.

Application of robots in the foundry rather than an assembly or machining division, has made it necessary to look for a means of locating target parts among a great deal of 'visual noise' on moving conveyor belts. For example, foundry finishing rooms do not have the luxury of having parts located in accurate fixtures on indexing transfer lines. Parts are scattered on the conveyor belts as they are discharged from tumble or shot blast cleaning machines or from grinding operations.

Finishing room personnel, standing alongside these belts, inspect the parts and segregate good from scrap by placing the scrap in a unique orientation. Downstream from the inspection, good parts are removed from the belt and placed on the shipping containers. The remaining parts go to scrap or rectify areas.

An automatic means was required to locate the good parts on the belt, recognise which of six castings each one was, remove it from the belt and place it into the proper shipping container. The parts come at an average line rate of 1,200 pieces/h with surges of up to 1,400 pieces/h.

Fig. 1. Castings are identified and unloaded to skips at up to 1,400 per hour by three Milacron T3 robots working with vision sensing at the GM foundry at St. Catherines, Ontario, Canada.

In 1979 a vision system was investigated – developed by the General Motors Research Laboratories and manufacturing development staff at Warren, Michigan – a system called 'Consight'. This system is capable of locating a part on a moving conveyor belt, and in near real time – about 200ms – transmit the pick-up point information to an industrial robot to enable it to pick up and dispose of the part into the proper container.

Previously it had been decided that vision was one solution to many of the material handling problems in the foundry, and in January 1981 work began with the Tech Centre to develop a fully functional system using 'Consight' and high technology robots capable of stationary base line tracking, sufficient computing capability to accept near real time changes in the user program.

At the outset of the project a major concern was the number of robots necessary to remove parts from the belt at the 1,400/h surge rate. One group felt that four machines would be needed while another estimated only two. The obvious compromise of three robots was made.

This uncertainty was typical of a recurring difficulty experienced in making reliable predictions of robot cycle times, since there are a great variety of programming approaches and compromises possible, and there is still not a good library of application experience available to draw upon.

The system consists of a vision structure straddling the process belt downstream from the inspection area, three robots sitting adjacent to the process belt and downstream from the vision structure, and a vision computer (Fig. 2).

Each robot has a single gripper capable of picking up any of six castings, and two shipping containers into which it may place parts (see Fig. 3). At the end of the process belt, 60ft. downstream from the vision structure, is a manual pick-off area and skid that collects any castings that require additional manual processing.

The belt runs at a 4° incline through the robot pick-up area so that the belt is high enough for parts to fall into the skid when the skid is at floor level. This dictates that the robots are able to track in two dimensions. In fact the robots being used are capable of tracking in three dimensions – a feature believed to be unique to this make of robot. Each robot is placed on a pedestal to maintain its proper height relative to the belt.

Fig. 2. Plan and overall view of the GM Canada robot vision system.

Fig. 3. Two of the T3 robots depositing rear end castings – in this case the castings are being held internally by the special gripper.

The Consight approach
Structured lighting

Many vision systems use an unstructured approach to lighting, in which the part to be viewed is back-lit through a translucent belt or surface of some kind, with the vision camera looking down to the back-lit scene. This approach is termed 'unstructured' because the method is simply to flood the underside of the work area with illumination.

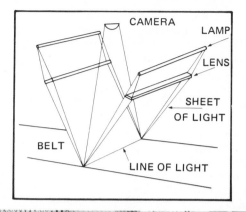

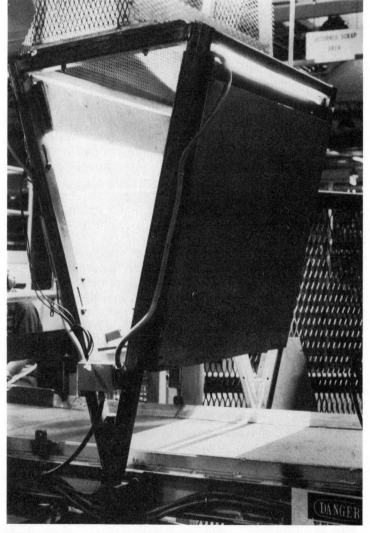

Fig. 4. Two 36in. linear sources project sheets of light at 15° on either side of the linear diode array camera which is mounted about 2 metres above the structure of the light housings.

The drawbacks of needing a translucent belt for back-lighting in a foundry environment become evident rapidly. Problems such as physical strength and loss of translucency due to parts scratching the surface preclude using this technology in the foundry. What was required was a technique which can see a black part sitting on a black belt.

Consight uses a structured approach to vision (Fig. 4). There are two linear light sources, each with a 36in./3¼kW lamp, and each illuminating a cylindrical lens 32in. long. The power and length of the light sources can vary between applications, depending on belt width.

Each lens focuses the light on to the surface of the belt, forming a line of illumination, approximately ⅛in. wide, across the width of the belt. By using adjustments built into the structure the two lines are focused and then superimposed on one another so as to give the appearance of only a single line of light.

About 12ft. above the belt – the exact height depends on the application – is located a 256 cell linear diode array camera which is focused on the line of light on the belt. A 256 cell array with a 36in. belt gives a resolution of 0.140in. across the belt.

Fig. 5 shows a general view of the lighting structure as seen from above the belt, with a part being illuminated by the sheets of light about ⅛in. wide. This is the view the camera sees.

Referring to Fig. 4, the included angle of the lighting structure light sources is 30°. Effectively this gives two sheets of light, each approaching the belt at an angle of 15° from the vertical. In Fig. 5 a part is shown illuminated. Note that the interruption of the sheets of light by the part causes the lines of light on its surface to shift, moving them out of the camera's field of view. This phenomenon can be seen in Fig. 6. The system works therefore, perhaps paradoxically, by responding when no light is present.

Fig. 7 shows why two sheets of light are used. If a single sheet is used the part's own shadow will give a false indication of the size and orientation of the part on the belt. The second sheet of light helps overcome this problem, as does the judicious choice of the included angle of the lighting structure.

There are significant trade-offs to be made with regard to the height of the part, minimum part spacing on the belt, and the included angle of the light sheets. For example, the tall parts will shade each other if they are too close together. To overcome this, the included angle of the light sheets could be reduced, but this then means that the parts must be higher to displace the light from the camera's field of view. The system sees the part as a silhouette in a white background, and the perimeter of the silhouette is the outline of the part at or above the minimum height.

Tracking – vision encoder and robot resolvers
At the take-up pulley of the belt is a structure housing a rotary optical encoder which yields data concerning belt position and velocity. The robot tracking resolvers which allow the robot to track the moving belt are also housed in this structure, and all these devices are driven by small chain and sprocket sets. The motive force is a rubber wheel which pinches the conveyor belt between itself and the take-up pulley of the belt.

The information from the encoder is fed into a vision preprocessor where it is massaged so as to provide, among other things, a scan signal to the camera once every 0.140in. of belt travel. This causes the camera to scan its array once, starting at element zero and ending at element 255. Since the transverse resolution is also 0.140 what effectively happens is to break the belt up into squares of 0.140in. sides. These cells are referred to as pixels – picture elements – and the state of each

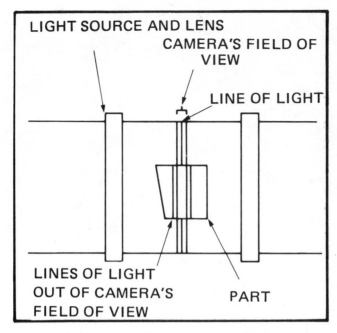

Fig. 5. Where a part passes through the line of light the beams from the two sources are deflected out of the camera's field of view.

element for each scan is stored in a buffer for processing. The vision computer removes the data for processing at the conclusion of its other tasks.

The three machines each receive belt position data from a dedicated rotary resolver at the belt tail pulley. The vision computer communicates with the robots via dedicated 9.6 kilobaud RS232 links and several discrete input/output lines as well.

System operation
Displacement of the sheet of light indicates when a part is present. This information is then used to locate and identify a part.

Visualise an object on the belt interrupting the sheets of light. It is known, by determining which cell on a particular camera scan was the first found to be dark, where one edge of the part is. The second edge is then located at the first cell to go back into light – assuming that there are no holes in the part. These holes are referred to as transitions, and there must always be an even number of these for every scan of the camera. The numerical difference between the cells at the transitions multiplied by the resolution yields the transverse dimension of the part. By storing successive scans as the part moves under the camera, part information is accumulated.

Once all the information for an objective has been accumulated the vision software constructs the image and generates the part area, maximum and minimum part dimensions, number of holes, second order moment and several other statistical characteristics. It then compares each of these factors with a previously taught library of parts, looking for a statistical fit. When the system is taught parts there is some control over the degree of fit needed, so that what is found is not a precise fit but more of a 'best fit'.

If a fit is found, then the system calculates the angular position of the part on the

Fig. 6. Views of parts passing through under the vision structure showing the two sheets of light diverging on the parts. On the left a rear end casting and on the right a calibration disc.

belt (Fig. 8) and locates the pick-up point for use by the robot. The robot is then flagged and the pick-up program sequence is transmitted to the robot. The robot then picks the part off the moving belt and disposes of it into the proper container.

The vision processing and object recognition portions of the program take a significant portion of the available computation time and are given top priority. This means that at times of heavy part flow, even though a casting may be within reach of a robot and the robot is ready, the system may be busy looking at parts and cannot service the robot immediately. During times of heavy part flow there may be a great deal of visual information present to be processed, causing the apparent speed of the system to degrade. Communication with the robot is a high priority task, and even at times of heavy part flow the delays are barely noticeable – under 500ms – unless you are familiar with the system.

A dispatching routine runs an algorithm which determines which castings will be picked by each robot. As parts are identified by the vision system their positional and pick-up co-ordinates are placed into a queue. As robots become available to pick up parts the information is removed from the queue and transmitted to the robot. Only those robots telling the vision system that they are available are considered. This availability is examined on a continuous basis, and the algorithm shifts the load to the available robots accordingly.

The system is always aware of which parts are within the reach of a robot, and if for some reason a robot cannot reach a part the missed part will be assigned to the next robot downstream.

The system will not allow different part types to be placed into the same container. A part can only be placed into an empty container or one which already contains parts of that type. If there is more than one part type on the belt, the dispatcher determines into which of each pair of containers for each robot the parts can be placed. The present system allows smooth running changes, by sorting two different castings, as the moulding line changes from one casting type to another.

The dispatcher routine also divides the belt into longitudinal slices, or windows. These may be different for each robot. There are also transverse windows for each robot which define the upstream and downstream reach limits, and estimate cycle and part acquisition times. This collection of times and dimensions allows the part flow through the system to be organised in a variety of ways, as the programmer chooses.

The dispatcher also performs a variety of housekeeping chores. The identity and quantity of parts in each shipping container is monitored. If a bin is full and another is not available the robot affected is stopped in cycle and the parts are directed to a robot with available bin space. The halted robot will restart when bin space is available.

If a robot malfunctions, it will flag the vision computer which will do an orderly shutdown of the data link to it and spread its part load among the functional machines. What is actually happening is that the available resources are always being used to maximum advantage. The system is very dynamic in its response to the availability of robots and the rate of part flow, the type of part on the belt and the availability of shipping containers.

In practice it has been found that two robots are able to handle the normal parts flow of 1,200/h. The system balance has been adjusted so that the upstream pair of machines attempt to remove 100% of the parts. The third robot sits waiting to be called upon during part rate surges or a malfunction of one of the upstream robots.

Teaching
There are three teaching tasks for the system – teaching of parts, calibration of the vision system, and teaching of the pick-up task.

Parts
To identify parts, the system refers to the library of statistical parts data. Teaching parts is the method by which this library is initially generated.

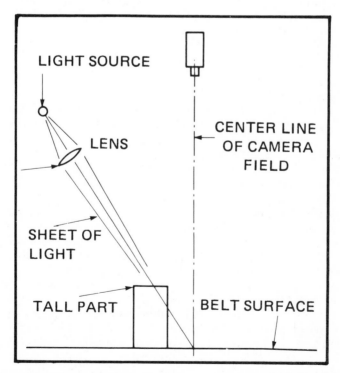

Fig. 7. If only one sheet of light were used a tall part would shadow the light before the part is under the camera.

When the vision system is placed into the 'teach' mode the system monitor will display an empty table of statistical measures. For each measure we have a maximum, minimum, average and standard deviation. A part to be taught is placed on the belt and transported under the vision structure by the belt, and the vision system computes all the part statistics.

The monitor then displays the data as gathered for the one part. This procedure is repeated about 50 times with the same part type in various orientations and transverse positions on the belt. It is advisable to use several different parts of the same type to account for manufacturing differences from part to part. This gives a good statistical basis for the library data.

At the conclusion of teaching, the information is stored in the parts library under a user-assignable name. This exercise is repeated for every unique part the system will be expected to recognise.

In many cases it is possible to assess which of the statistical measures have the greatest effect in discerning one part from another, and these may then be used exclusively. The remaining statistical measures may, in effect, be turned off by making their maximum and minimum values equal the maximum and minimum real numbers that the vision computer can manipulate.

Calibration
Calibration provides the vision system with the necessary data to locate the origin of the vision coordinate system within the robot co-ordinate system, and to account for rotational misalignment between the two co-ordinate systems.

Previously taught circular calibration discs are run under the vision structure to within reach of the robot. The robot gripper is then moved to the centre of the discs and its co-ordinates are entered into the vision computer. The vision computer uses these data to calculate the necessary rotational and positional offsets.

This calibration must be done as accurately as possible, since a calibration error would make it more difficult to pick up a part successfully from the belt.

Pick-up
After system calibration, a part of each type to be handled by the robot must be run under the vision system to a point within reach of each robot in turn. At each robot the gripper is moved to the desired pick-up point on the part. The robot co-ordinates are then entered into the vision computer, where the pick-up offset is calculated.

This being the last operation taught, any errors in vision calibration or the robot tracking calibration will show up here as large errors or as inconsistencies from location to location on the process belt.

Operating experience
The system described is felt to be a significant step forward in material handling technology. The actual results indicate a rate in the region of 600 parts/h per robot without undue mechanical stress. There are, of course, many improvements and enhancements that can be made to this system to improve its flexibility. Improvements such as recognising touching parts, or parts with the sprue still attached are the obvious next developmental steps for Consight.

It has proved that the concept of multi-robot vision is applicable and practical in the harsh environment of the foundry, and should therefore be suitable for almost any manufacturing environment. Consight can be adapted to an application with little alteration to the existing process equipment.

No major problems in installing and running the system have been encountered, but the following are some of the minor difficulties that had to be overcome.

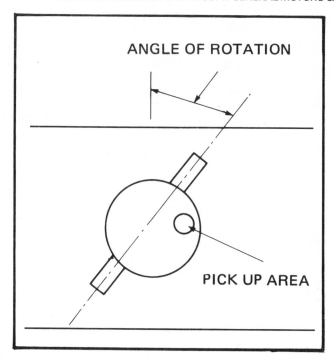

Fig. 8. The angle of rotation of the part is a crucial factor determined by the vision system.

Because Consight uses reflected instead of transmitted light a difficulty can arise with stray reflections. In the current process castings are shot blasted to clean them, and this operation takes place upstream of the vision system. Castings which are to be further manually processed or have not been properly deflashed are removed from the belt by inspectors for recleaning.

This double cleaning results in the surface of the casting taking on a sheen which when combined with casting surface geometry yields highly reflective pockets which are interpreted by the vision system as holes in the casting. The true number of holes is between zero and seven, depending on the casting type, but modifications to the parts library had to be made, to increase the maximum number of holes expected.

The lenses are made of a cast acrylic with transparency optimised in the infrared region (700-1,100nm). They are 32in. long, 2in. wide and a maximum of 1in. thick. The light sources, being 3.5kW each, cause the lenses and supporting structure to become very warm. A structure of this size had not been built before and was designed initially with the lenses only restrained at their ends. They distorted into a 'C' shape, resulting in a defocusing of the lines of light, so it was necessary to modify the lens supports to inhibit this thermal distortion.

Dust collecting on the lenses has, surprisingly, not posed a problem. The lenses are blown off every few days, and the layer of fine dust which collects on them has never degraded the lighting sufficiently to blind the system.

When the system was started, it began with a new process belt which, because of its method of manufacture, lay concave on the conveyor bed plate. The edges were initially ¾in. above the belt centre. This deflected the line of light and made Consight believe that a long narrow part was at the belt edges, and made one of the vision tasks fail as the part was excessively long. Small clips have been attached to the bed plate to restrain both vertical and horizontal movement of the belt in the illuminated area.

The vision task loads a buffer with camera data upon receipt of each belt interrupt. The impact of castings falling from the grinder exit on to the belt causes momentary and rapid changes in belt speed. If these changes are excessive, information is missed and a part is not recognised. Modifications at the grinder exits have reduced the severity of this problem.

Any optical system must of course have a limited field of view. The difficulty is that Consight cannot process part data if a portion of a part lies beyond the field of view. Two of the castings have large tabs which occasionally overhang the belt edge and then are not recognised. Expansions to the field of view are currently being experimented with, but this entails a corresponding reduction in resolution.

The problem of touching parts, which cannot be distinguished by Consight, has already been mentioned. An enhancement may become available if demand warrants it, to eliminate this problem. In this application the parts are asymmetrical and tall, so that in some orientations they may be placed closer to each other than in others before part shadowing occurs. The inspectors upstream of the vision system are aware of this and attempt to ensure adequate spacing – $\frac{3}{4}$ to $1\frac{1}{4}$in. depending upon orientation.

During periods of heavy parts flow it is not possible to ensure that the minimum spacing is maintained and the occasional part is missed. No modifications are anticipated since the number of castings missed is not significant.

PATTERN RECOGNITION IN THE FACTORY: AN EXAMPLE

Jean-Paul Hermann et la section 'capteurs', Direction des Techniques Avancées en Automatismes, Régie Nationale des Usines Renault, France

Presented at the 12th International Symposium on Industrial Robots, 9-11 June 1982, Paris, France

Abstract

A practical application of pattern recognition installed in the Renault factory at Cléon (France) is described. This concerns the automatic depalletisation of crankshafts by an industrial robot. The need for several types of sensors is stressed and the problems encountered during the installation discussed. The sensors used are: a TV camera with appropriate illumination and interface systems, a sonar carried by the robot, two photocells and several inductive sensors. The strategy of the management is of utmost importance to insure a high reliability of the system.

The problem

Let us first give a general outlook of the problem.

History

Some time ago we were asked to design a system to automate the loading of a transfer line which machines crankshafts. This was previously done manually, the worker having to carry 12kg crankshafts at a rate of three per minute. This was a tedious and tiring job so that automation was highly desirable.

Crankshafts are manufactured and arranged in pallets in the foundry at Douvrain, more than 300km away from Cléon. During shipment, some disorder unavoidably occurs on the pallets, so that some kind of pattern recognition is required for a safe handling of the workpieces.

The data

The pallets have a standard size of 1400mm × 1100mm and each carries 108 crankshafts, arranged as six stacks, each stack being made of nine sheets of two crankshafts (see Fig. 1). As seen from above, the pallet looks dark with the crankshafts on top looking shiny. Each stack is prevented from slipping and touching its neighbours by means of a special plastic form fitted on top of the wooden pallet; the crankshafts interlock on each stack; this helps them to stay more or less in place.

As can be seen, the system is a three-dimensional semi-ordered one. As said before, it is not possible to keep a perfect order because of a shaky shipment and by the same token, it would not be practical to deliver the crankshafts in bins because this might damage them and they would be entangled in such a way that even a man could not manage such a mess. Thus the arrangement of the workpieces into stacks is a good choice that insures both dense packing and fairly good accessibility to the crankshafts. This choice has been made years ago, long before the age of robots.

Pattern recognition

The problem to be solved is quite different from the identification and localisation of workpieces on a conveyor belt, which we demonstrated years ago, along with many other workers. The main differences (and also the main difficulties) are the following:

☐ The problem is a three-dimensional one, since the height Z of the crankshaft has to be determined as well as its position X, Y in the horizontal plane and its orientation θ. Thus four parameters are required prior to each grasp: X, Y, Z and θ. The arrangement in sheets eliminates the need for two more angles.

☐ The contour of the workpiece does not appear clearly since a camera cannot help but see the crankshafts located underneath (see Fig. 2).

☐ The system has to work in a real industrial environment, two shifts, in the dirt and under any external lighting conditions with a success rate higher than 99.9%, that is no more than one call for help per shift.

The robot and its mechanical environment

The robot chosen for this application is a standard hydraulic Renault-Acma V80, equipped with the now commercially available V5 control system. This system is a

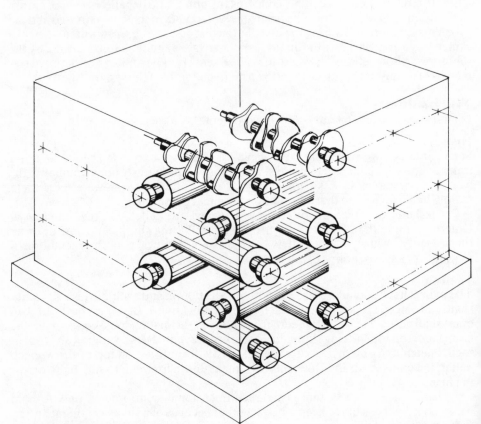

Fig. 1. General picture of a pallet.

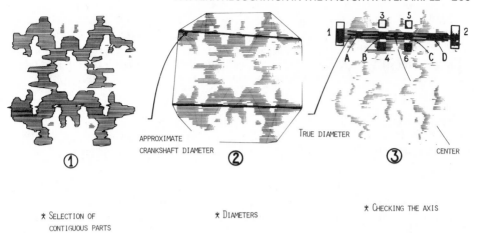

APPROXIMATE
CRANKSHAFT DIAMETER

TRUE DIAMETER

CENTER

① ② ③

★ SELECTION OF
CONTIGUOUS PARTS

★ DIAMETERS

★ CHECKING THE AXIS

Fig. 2. Crankshaft pattern recognition.

highly sophisticated one; among many other features, the most important for the present application are the possibility of connecting a pattern recognition system, a microprocessor driven sonar and also the ability to accurately alter its trajectory according to the data sent by the sensors.

The end effector

The use of an electromagnet was precluded for the handling of the crankshafts to avoid any magnetisation that would attract chips during the machining process. Thus we designed a special gripper with two hydraulically actuated 'fingers'. When the gripper closes, the first finger that touches the workpiece stops and the full strength is not delivered until the second finger has touched the workpiece also. This is done by a purely mechanical system and gives the advantage of leaving some tolerance to the positioning of the gripper, and does not shake the stack in case of a slightly inaccurate grasp.

For safe grasping, the gripper must be positioned within ± 15mm and $\pm 2°$ of its actual position (a crankshaft is 50cm long and 20cm wide). This sets the limit of the error that can be tolerated from the sensors and the pattern recognition system. As a matter of fact, this tolerance is much less than the possible fluctuations of the pallet around its nominal place (more than 10cm).

The entrance of the transfer line

The transfer line entrance is a 'caterpillar' conveyor that is also used as a buffer to feed the machine during stops of the robot (essential for when a worker comes in to change the empty pallet for a full one). The robot has to lay the crankshaft down on a cradle made of two V-shaped grooves; then a lifting system is used to push the workpiece up and drop it down on the conveyor. The design of this system is standard and leaves a rather small tolerance (± 5mm) for the robot. This has a nasty consequence, which is that on one hand the robot can grasp the crankshaft with a longitudinal error of 15mm but on the other hand it must lay it down within 5mm. In other words, an additional sensor is needed to determine the actual position of the workpiece inside the gripper.

General outlook of the system (see Figs. 3 and 4)

The robot has two pallets at hand, one on each side and it picks from one of them till it is empty; then it warns the operator and starts working on the other side without wasting any time waiting. Due to safety regulations, the whole robot is fenced in and it must stop during a pallet change. The robot has a span of 2m, which is fully used in this application. Its accuracy under geometrical control (not to be confused with repeatability) is ± 1mm. So full advantage is actually taken of the characteristics of the robot. The 10cm left on the sides of the pallet are sufficient for the fork truck driver to put it in place easily.

The sensors

We shall now describe in some detail the sensors we used for this application, keeping in mind the need for them to work in an industrial environment.

The TV camera

As mentioned before, the aim is to see 'shiny' crankshafts on a 'dark' background. This is not so simple since a shiny crankshaft may get dull when slightly rusted and a dark background may become bright if some oil or grease has been accidently poured on it.

The requirements are the following:
○ The camera contrast must be good,
○ the camera has to adapt to varying lighting conditions and varying workpiece reflectivities,
○ the resolution (pixel size) must be no more than 5mm,
○ the lighting must be appropriate.

Our camera choice is a Thomson Newvicon TV camera, specially modified according to our specifications. The advantages of this camera are: a better contrast ($\gamma = 1$) than a vidicon ($\gamma = 0.6$) and an excellent resistance to blooming. The

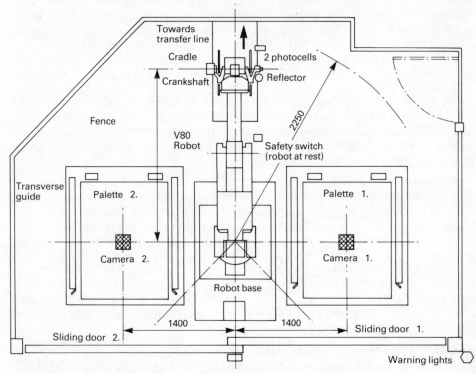

Fig. 3. Layout of robot depalletisation/machine loading cell.

Fig. 4. Renault-Acma V80 transfers crankshaft from pallet to input conveyor of transfer line.

diaphragm of the lens is self-adjusting so that the peak video signal remains constant. The camera is synchronised with the computer by means of a special cable.

The field of view contains 256×382 pixels. For a camera fastened at a height of 5.6m above the pallet and equipped with a 25mm focal length lens, this corresponds to a pixel size of 4mm. The camera is enclosed in a sealed box attached to the building frame. Some reference marks are drawn on the floor to allow for periodic checking of the camera calibration.

The illumination is provided by a circle of lights around the optical axis of each camera; this system is designed so that no ray can travel from the lamps to the lens after reflecting on a horizontal plane (which can be for instance the plastic on top of the wooden pallet, or the floor or an oil splatter). Some unpredictable lighting may come from the windows of the building but this is no problem because these rays are slanted and the diaphragm self-adjusts.

There is of course one such system on top of each pallet.

The algorithm we created for this application makes use of the size of the crankshaft so that it is necessary to know at what height the crankshaft is located before the computer locates it in the horizontal plane. This height may vary between 20cm and 1m from the floor depending on the degree of depalletisation. This corresponds to a variation of 14% in the size of the picture and we must take care of it. This perspective effect could be reduced if the camera was located higher but this was not practical in this application because of a low ceiling. This effect leads to a severe paradox to be discussed later.

The sonar

A sonar is used to determine the heights of the workpieces above the floor. Ours is a standard Polaroid transducer located inside the gripper with its electronics; we modified the timing circuitry to improve the accuracy of the measurements. The specifications we determined by reflection on a normal plane are:
○ useful range: between 25cm and 4m,
○ accuracy: ±0.2%,
○ linearity: perfect within the above mentioned accuracy,
○ directivity: emission cone narrower than 10° full angle.

The transducers work around 60kHz, its diameter is 5cm. Since it is not possible to increase either the frequency (due to sound absorption in air) or the diameter (due to size and modal problems) and since this directivity is essentially limited by the diffraction, it does not appear possible to improve any characteristic unless the whole workshop is flooded in water which is hardly acceptable.

We attached a special foam sheet to the transducer, so as to protect it from the ambient dirt; this does not alter its properties.

The problem inherent to the sonar is its limited directivity; thus it has to be carried by the robot and precisely located at the correct height above the target which is far from being simple.

This closes the vicious circle:
○ the robot needs the camera to know where to go,
○ the camera needs the sonar because it requires the knowledge of the height of the workpiece,
○ the sonar needs the robot so it can be carried along.

To get out of this loop we operate by successive approximations as described below.

Operational procedure

To describe the procedure, let's start from scratch: on a Monday morning, power is put on, all memories are empty. Successive operations are as follows:

□ **Calibration of the sonar**
Since the speed of sound is temperature dependent, one has to calibrate the sonar. This is easily done by aiming it to the floor and making three measurements at three different heights. The computer checks the linearity and calculates the relevant coefficients. Systematically, all sonar measurements are made three times at each height and recorded only if the fluctuation is less than 3mm. This avoids trouble in case the transducer detects some of the ambient noise; actually the

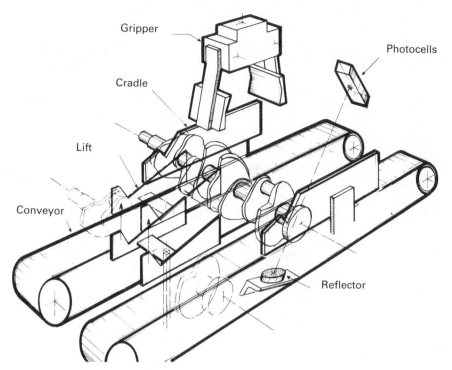

Fig. 5. Arrangement for depositing crankshafts onto input conveyor.

directivity of the transducer brings a major contribution to the elimination of spurious acoustic signals.

☐ **Determination of the lowest pallet**

The sonar is then carried 3m above the nominal position of each pallet to determine which one is the lowest: depalletisation will start with that one.

☐ **Determination of low stacks**

The previous measurement could only give a global indication about the height of the pallet. Now we want to know if for some reason one stack might be lower than its neighbours (this would happen if someone has taken a sample for quality inspection). In that case, we do not allow a grasp on that stack, otherwise the robot might hit the other stacks during its motion.

The determination of the height is made by carrying the sonar 70cm above the previously determined height of the pallet. The localisation in the horizontal plane is assumed to be the nominal one. At a height of 70cm the field of the sonar has a diameter of 15cm, which is more than the dispersion of the position and more than the diameter of the hole left between the crankshafts. Thus this measurement gives the height of the stack with an accuracy of ± 5cm (remember that the accuracy of 0.2% is valid only for a plane target). This accuracy is not good enough for grasping but it is sufficient for the pattern recognition algorithm.

Now the robot withdraws to allow the camera to analyse the scene.

☐ **Stack analysis**

Time has come for a first crude pattern recognition. The camera analyses the scene after binarisation with an adjustable threshold. The picture of a stack looks like Fig.

2. It is obvious that due to the lack of depth sense, even a human observer can have trouble telling which ones of the four crankshafts are located on top. As a matter of fact, it does happen that sometimes a crankshaft located on top is more blurred than the one located underneath (in manual operation, the worker has an oblique view that makes things easier for him).

To overcome this difficulty we designed a special algorithm that identifies the stack as a whole and locates its centre. This is done for the six stacks, except if one was found to be too low, in which case operation on that one would be delayed.

□ Sonar investigation
Only the sonar can give that sense of depth; thus we carry it again over the stacks but as low as possible (30cm) and aim it at four successive points and record either a succession like: bump-dip-bump-dip or dip-bump-dip-bump or dip-bump-dip-superdip, which allows us to tell respectively that we have: two crankshafts along the X axis (or nearly so) or two crankshafts along Y axis, or one crankshaft and the Y axis and one missing.

Notice that at a height of 30cm, the field of view has a diameter of 6cm and thus the sonar beam is able to propagate between edges 8cm apart and will bounce off lower crankshafts. Full advantage is taken of the directivity of this kind of sonar.

□ Crankshaft pattern recognition
At that stage the computer has gathered all the data it requires:
- ○ the precise ($\pm$ 5cm) height of the crankshaft,
- ○ the approximate ($\pm$ 10°) orientation of the crankshaft,
- ○ the approximate ($\pm$ 5cm) position of the centre of the stack.

A square picture is taken centred around the centre which is processed as follows:
- ○ identification of contiguous zones of the pictures,
- ○ calculation of the convex circumscribed polygon,
- ○ location of the diameters of this polygon,
- ○ sorting of the diameters according to their lengths,
- ○ testing of the succession of black and white zones along the diameter (see Fig. 2).

□ Grasping
The robot is sent to the precise location of the crankshaft but before seizing it the robot makes an ultimate sonar measurement to have a very accurate ($\pm$ 2mm) value for Z. It is crucial that the height be correct. Otherwise the robot could crush the sonar or lose the workpiece, jamming the whole stack.

□ Laying down
As stated earlier, the robot must lay the workpiece down on two V-shaped grooves. There is an additional difficulty, which is the fact that sometimes the pattern recognition system is in error by 180°, which does not affect the grasping but is not acceptable by the transfer line (the two ends of the crankshaft are not identical, one is 'fat', the other is 'slim'). Thus we have installed two photocells and move the robot gripper till one beam is obscured. This gives an important indication as to the location of the crankshaft inside the gripper and we use it to correct the trajectory, reaching the $\pm$ 5mm accuracy required for a proper laying down.

The second beam is obscured by the fat end if it happens to be at that end, which is wrong: in which case the crankshaft is turned over (see Fig. 5).

It is worthwhile saying that the whole procedure 1 to 8 is not repeated for each grasp. The following occurs:
- ○ every day 1 to 8,
- ○ every pallet 3 to 8,
- ○ every grasp 6 to 8.

So the procedure is done faster than it can be written! The extra operations 3 to 5 take about 30 seconds, grasping is done at a rate of three per minute.

The pattern recognition is fast: it takes less than two seconds for a pair of crankshafts. This is due to the use of a special proprietary multiprocessor calculator that synchronises the camera and gathers the data.

It is not the purpose of this paper to describe in too many details the management of the whole installation. Let us just mention the use of several inductive sensors and switches which are used to detect:

○ the filling of the buffer so as to prevent the robot from laying down a workpiece on a full conveyor,
○ wrong positioning of the crankshaft on its cradle: the lifting system must be stopped to prevent jamming,
○ a call from the fork truck driver who wants the robot to be stopped to install a new pallet; this may happen any time,
○ an unwanted move of the robot during the presence of a man inside the fence; this safety switch avoids disabling the servo-mechanisms too often.

Conclusion

This paper is intended to stress the complications that arise when one wants to transfer a technique from the laboratory to the workshop. Problems of accuracy, control of the environment, safety, etc., take on a considerable importance that has too often been overlooked in many publications. Even a very low artificial intelligence technique is a very intricate subject. It is not possible in a workshop to finesse with any mechanical or environmental detail; this is why there are so few intelligent handling systems in factories, despite a relatively low hardware added value. The system described above is actually one of the few industrially operating.

A considerble effort will be necessary in the future to broaden handling techniques by robots, and even more so in assembly, where something like the introduction of 'artificial dexterity' will be needed.

Acknowledgements

This work is the result of the collaboration of many people, among whom were: J. Bach, J. Chabrol, J. Chevalier, T. Craplet, J. L. Fontaine, D. Huet, I. Lacrouts-Cazenave, J. LeRouzo, G. Michard, P. Pardo, J. R. Passemard, V. Paternoster, M. C. Petit, D. Salafia and A. Rousseau.

A CCTV CAMERA-CONTROLLED PAINTER

E. Johnston, Haden Drysys International, UK

Presented at a colloquium 'Industrial applications of robot vision', 24 November 1981, Institute of Electrical Engineers, London

Introduction

The machine concerned is a reciprocating spray gun. This machine consists of a spray gun that can be switched on and off and is moved back and forth over a fixed length path, whilst the product to be painted is passed in front of this reciprocating spray gun. Conventionally the control of the spray gun was achieved by toggle switches mounted on a switch bar along the path travelled by the spray gun. This method was suitable for long production runs where there was no need to modify the spraying stroke very often.

But eleven years ago the need was present for a more complete control of the spray gun operation. On the one hand reciprocating spray guns were being used to paint the sides of automobile bodies on a production line, where the required pattern of spray gun on and off varied stroke by stroke along the length of the body, and on the other hand reciprocating spray guns were beginning to be used for shorter production runs where the spray gun stroke pattern was needed to be changed quickly and often. In the first case the method of control adopted was to pre-program the required stroke pattern for each particular body that would arrive at the painting station and store these programs in memory to be called up and used as required. In the second case multiple cam systems were developed to provide a means of quickly changing from one stroke pattern to another.

At that time it seemed that a more comprehensive method of control for the reciprocating spray gun would be to use a CCTV camera to provide information on the product passing in front of the spray gun and use this information to switch the spray gun on and off successively as it passed across the product. In this way the need to pre-program for any particular shape that might be presented for painting would be eliminated and so the need to change cams for different stroke patterns would be removed. In effect the CCTV camera controller system would act as an automatic self-setting electronic cam. This system was called 'Videospray' and after five years in the development laboratory was ready for industrial use.

Design philosophy

The original concept was that the camera would view the product directly in front of the spray gun and control the spray gun 'on the fly', so to speak. Unfortunately a curious effect took place on testing the system. As the product appeared in front of the spray gun the controller opened the spray gun to paint the product. However, the camera now saw a dense cloud of paint that obliterated its view of the product and it carried on painting the paint cloud – an unusual case of positive feedback. At the same time it became apparent that this kind of control would only be of use when one spray gun was being used and also the spray gun would need to be seen by the CCTV camera.

Subsequently the philosophy was established of waiting until the reciprocating spray gun was at the end of its stroke and changing its direction of motion. At this point, because the motion is non-linear, the spray gun would always be closed; thus the CCTV camera could look at the product along the path that the spray gun would be following on its next stroke, make a number of decisions as to when to switch on and when to switch off and put these decisions into memory. Subsequently as the spray gun travelled along this path it would recall these decisions and act upon them. Having so generated a stroke pattern and placed it in memory, it then became possible to use it more than once. For instance, if more than one spray gun was mounted on the reciprocator, then as the product eventually moved in front of the second and subsequent spray guns the same stroke pattern could be used but suitably delayed.

One of the first decisions made was to use a reflected light system rather than a silhouette system. For the first idea of painting 'on the fly' it was not possible to use a silhouette system because the background of the camera's field of view would always be the spray booth and as such could not reasonably be uniformly illuminated. But for the final development, where stroke patterns were generated and put into memory, it became possible to use a silhouette system. Unfortunately at the time that this development was taking place, during 1973/74, static memory was still expensive, i.e. £100 per kilo bit, and this was intended to be a low cost system; so the design decision was to stay with a reflected light system. This was a fortunate decision, since the result was a very flexible situation when it came to the installation of the equipment into the various factory environments.

Obviously a device like a spray gun does not have the instantaneous response of an electronic signal, so some form of compensation for system delays had to be incorporated. The principal delay was the opening of the spray gun, that is the withdrawing of the needle by a compressed air cylinder, and the subsequent build-up of the atomised spray fan. Throughout this paper it must be borne in mind that with a paint spray system we are dealing with a hazardous situation. Hazardous, that is, as defined by BASEEFA 0298 6211; in this case, conventionally, control was achieved by compressed air signals.

In order that installation would be as simple as possible, it was decided to provide a small vision monitor fixed 'piggy back' fashion to the camera and usually called an electronic viewfinder. This enabled the required scene to be selected very quickly and the camera locked in place. Subsequently the monitor became a useful screen on which to display various video-generated markers and signals, thus dispensing with the need for an oscilloscope and voltmeter when setting up on site.

One other item to be decided at this stage was the precision of operation. Typically we were considering a stroke length around 2m and a spray fan with an elliptical cross section at the product surface of 150 to 250mm major axis and 50 to 100mm minor axis. Thus a basic accuracy of 20mm seemed both appropriate and feasible.

Summarising, we were aiming to provide a control system that would look at the product to be painted as it passed along towards a reciprocating spray gun. It would then make a series of decisions regarding the product and switch one or more spray guns on and off as they passed across the surface of the product, to an accuracy of

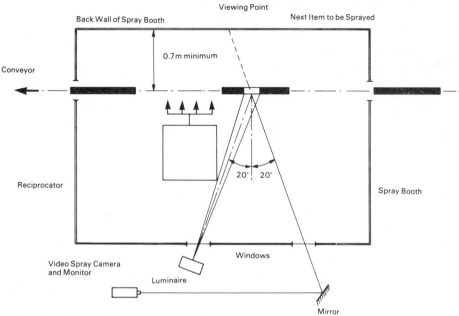

a. *Plan of typical installation.*

b. *Rear view of an installation showing beam of light on object and camera monitor.*

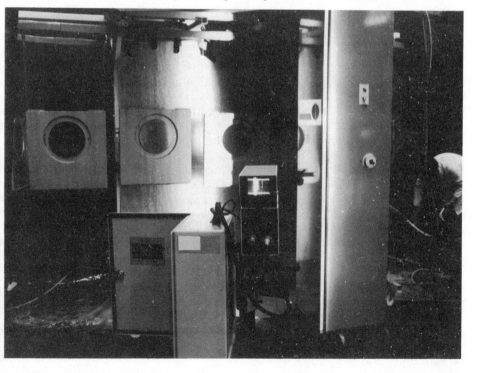

Fig. 1. *Typical installation.*

20mm, the object being to minimise the amount of paint and so provide a more efficient use of the paint material.

Design description

In such a short paper it is not possible to give a detailed design description. for this reason there has not been included any form of block diagram. Instead the major design points only will be described. One other reason for not dwelling too long over detailed description is that the design suffers from its historical age. In other words it is hardware oriented and there is no microprocessor.

Sources of Information

1. A CCTV camera provided information on the product to be painted.
2. A pulse generator A, fitted to the reciprocating mechanism, provided pulses at the rate of 1 pulse per 20mm or thereabouts.
3. A pulse generator B, fitted to the reciprocating mechanism, provided a pulse whilst the spray gun was passing through bottom dead centre.
4. A contact that was normally closed when the conveyor, carrying the product, was in motion and open when the conveyor was stopped.

In order that the system should be inherently stable in operation it was essential that a good contrast was obtained between the product to be painted and its background. Since the product to be painted could be of any colour and any material, from dead flat matt black to polished highly reflective aluminium, and the background could be the same, it became necessary to devise a system that could detect black objects against a black background and also bright objects against bright background. This was clearly not possible so the following strategy was adopted, with reference to Fig.1.

Arbitrarily it was decided that the product would always be made to appear white and the background would always be black. To achieve this a very powerful light source was arranged to illuminate the product along the line that could be considered as a possible path of the reciprocating spray gun. This line of light was positioned a short distance ahead of the actual path followed by the first spray gun. The aim was to achieve 1000 lux on the product but generally anything higher than 700 lux was sufficient. At the same time it was arranged to create a deep shadow of less than 100 lux over the background. In other words, by shielding it was possible to exclude or prevent light from falling on the back of the spray booth. Satisfactory operation was achieved with values less than 200 lux. In plan view, the line of light was made to impinge on the product at an angle of incidence of 20°. It was then arranged for the CCTV camera to view the scene along the line of the angle of reflection. The result of this arrangement was that the camera saw the product glowing whitely against a black background. Since we were only interested in that part of the product that would be painted in one stroke of the spray gun, only that part was illuminated. So the scene viewed on the monitor was generally a normal picture with the product glowing whitely in mid-screen against a black background. Since this last part was not a normal scene, it needed care when viewed or interpreted by human viewers.

Having established a good contrast between product and background, it was then necessary to detect this difference and process it. Because a good contrast had been established a simple binary slicer circuit was used. The slicing level was set by a 20-turn potentiometer whilst viewing the effect of the slicing on the monitor. In other words as the white to grey levels are detected and converted to full binary one values the result was mixed with the normal video and displayed on the monitor. The effect was for the white areas of the scene to 'bloom' and as the slicing level was progressively lowered so the grey areas were caused to 'bloom'. A satisfactory slicing level was achieved when the product shape as illuminated by the line of light was sharply outlined within the line of light. This binary sliced result was termed video squared.

It was then necessary to select the information that would be needed for control and reject all redundant information. To do this a window was generated. The window was meant to represent one painting stroke of the spray gun across the product and when mixed with the video and displayed on the monitor appeared as a vertical white stripe. Subsequently the width of the white stripe was adjusted by means of a 20-turn potentiometer to equate with the actual spray fan width of the spray gun. Then the position of the window was adjusted by means of another 20-turn potentiometer so that it was centred on the line of light illuminating the product. Thus when the window waveform was used to gate the video squared waveform the result was, line by line of the video picture, a series of decisions about the product to be painted on that line or not.

One scan of a standard TV picture takes $312\frac{1}{2}$ lines; of these the first 19 or 20 are not available since they occur in the frame blanking period – this leaves 293 lines yielding information. Our requirement was to control to 20mm in a 2m stroke, or 1% resolution. By using every other line, only the information on 146 lines was used. At this point it was necessary to look at the memory IC's that were available; a convenient size was a 128 bit memory. This would enable the spray gun stroke to be divided into 128 strips and an individual decision to be made for each strip so that spray gun control to better than 1% was achieved. It did mean, however, that some means of selecting in which 128 lines of the 146 lines were to be used had to be provided. To do this it was first necessary to count the lines as they were generated, so another binary slicer was used, this time preset to catch the sync pulses that were present at the beginning of each line. These sync pulses so generated were divided by two then counted with a 7-bit binary counter. When the count reached 127 the counter was disenabled and the count held at 127 until reset. The reset pulse was generated by feeding sync pulses to retriggerable monostable. When the sync pulses ceased, during the frame blanking period, the monostable recovered and thus produced a reset pulse.

When we considered the spray gun motion we established three zones. They were: a central zone, which was the zone of linear motion, and two end zones where the spray gun changed its direction of motion and the motion was non-linear. It was further established that painting would only take place in the central zone and would be inhibited in the two end zones. It was then necessary to indicate the boundaries of these zones on the monitor screen so that they could be aligned with the physical scene. It was decided that the 128 lines should be aligned to span the central linear zone, i.e. the painting zone. From this it followed that the TV line with strip number 0 marked one boundary and the line with strip number 127 marked the other boundary. These lines were therefore brightened and mixed with the video waveform to produce two white horizontal lines on the monitor screen. These lines were designated upper limit marker and lower limit marker. As described so far there was no means of adjusting the position of these two lines, so a 6-bit digital delay was inserted between the divide by two circuit and the 7-bit binary counter in the sync pulse line. This digital delay was made adjustable with a 16 position switch and designated picture shift control, thus enabling the two marker lines to be stepped up or down on the screen to be aligned accurately with the actual motion limits of the reciprocating spray gun. Before leaving the subject of the limit markers it must be pointed out that so far the assumption has been that the camera has been placed in the optimum position for the apparent vertical motion of the spray gun to just fill the monitor screen. It was considered probable that this would not always be the case and that sometimes the limit markers would have to be moved inwards individually until they marked the limits of linear motion of the spray gun, so two additional 16 position switches were installed, one to adjust the upper limit marker over the upper 25% of the screen, the other to adjust the lower limit marker over the lower 25% of the screen. The action of these two switches is a straightforward number gating function to produce line bright-up on the required lines. The names of these two switches are self-evident.

Summarising, so far we have established a scale of 128 strips, aligned and proportioned them to fit the viewed scene, produced a window to derive the product-to-be-painted information and aligned and proportioned it to match the viewed scene.

By correlating the window gated video squared waveform with the line counter it was possible to derive a unique spray gun on or spray gun off instruction for each of the numbered strips. A memory location was made available for each of these strips and the 1 bit instruction stored therein.

The decision whether to write into memory or read from memory was obtained from a knowledge of where the spray gun was at any particular moment. This knowledge was derived from the pulses supplied by the two inductive proximity sensors fitted to the reciprocator mechanism. Pulse generator A supplied a pulse for every 20mm of spray gun motion and these pulses were fed to a phase locked loop which multiplied the number of pulses produced by 256. Subsequently these were divided by an adjustable three digit decimal divider. This divider enabled reciprocators of different spray gun path lengths to be accommodated. (N.B. Reciprocators with from 1.2m to 4.2m path length could be supplied). Another way to consider the effect or use of this three digit divider was as a scaling device for the fine trimming of the scaling of the physical image to the electronic image. The result was that the output from the divider produced 256 pulses for a complete cycle, i.e. 128 pulses for up stroke and 128 pulses for down stroke.

These 128 pulses then had to be aligned with the central zone of the reciprocator motion. To do this the output from the divider was counted by a 7-bit binary up/down counter and its output compared with the output from the line counter of the CCTV camera. When coincidence occurred a bright-up was produced and used to mark that particular TV line. The effect was to see a white horizontal line moving up and down the TV monitor screen.

The pulse from pulse generator B that occurred once per cycle was used to reset the counter, so that errors in scaling did not accumulate beyond one up stroke and one down stroke. Since generally some part of the reciprocating spray gun or spray gun holder was always seen by the CCTV camera, it was a simple matter to adjust the three adjustable decade switches of the three digit divider until the white horizontal line moved up and down the monitor screen in synchronism with the spray gun or spray gun holder. The use of a phase locked loop was justified on the basis that there was usually a 5% speed difference between up stroke and down stroke, depending upon spray gun weight and also the customer must be free to adjust the reciprocator speed without having to readjust the electronics.

At this point the white horizontal line was not necessarily coincident with the spray gun as it moved across the screen, but it should have been travelling at the same speed. To bring the two into coincidence it was necessary to add a few pulses' worth of delay at either end of the traverse to account for the two end zones of non-linear motion. For this a 4-bit digital delay was inserted between the three digit divider and the 7-bit up/down counter. A sixteen position switch was then adjusted to align the white line and the spray gun during the up or outward stroke and a second sixteen position switch was used to align them during the return stroke. By this means the position of the spray gun or guns was known at any instant and thus it was possible to state that when the spray gun was outside the central zone, i.e. it was in an end zone, then during the time data would be written into memory and when the spray gun was in the central zone data would be read from memory. So for each painting stroke of the reciprocator we now had a sequence of write into memory, read from memory for as long as the reciprocator was reciprocating. Another quickly learned lesson was not to be too clever and use the camera to switch on the reciprocator when a product arrived at the paint spray booth and then switch off as it left, because a man going into the booth to clean the spray nozzles of a stationary reciprocator was likely to cause start-up as he was seen as a possible product to be painted. It was far safer to make the reciprocator on/off control a solely manual function.

It has been stated previously that it would be necessary to store more than one stroke's worth of data, since the viewing point was slightly ahead of the painting point along the direction of the product conveyor. It was decided that sixteen strokes would be a convenient number of strokes to store, since this typically represented about 1.5m of conveyor travel and it was unlikely that the spray guns mounted on the reciprocator would be spread out more than this, i.e. the spray gun furthest from the viewing point should not be more than sixteen strokes away from the viewing point because it would then be outside the memory range of the system. Sixteen strokes was also convenient because the particular stroke delay needed by spray gun could then easily be selected by a sixteen position switch.

An analysis of early contracts showed that individual control of four spray guns would cover all of the installations to that date. So four sixteen position switches were provided for the individual control of the four spray guns that could possibly be required. Typically only two have been used.

One problem remained. What would happen if the conveyor stopped? Unless something was done to stop it the spray guns would continue to spray according to the memory contents and the memory would gradually fill up with whatever condition was present at the viewing point, because it was safer to make the reciprocator start/stop a manual operation. For this problem we used a contact from the conveyor motor controller such that when running the contact was closed, when stopped the contact was open. This contact was used to inhibit the write/read operation and the spray gun operation for as long as the conveyor was stopped. In this manner the memory contents stayed in step with the conveyor contents despite the continuously running reciprocator.

Returning now to an earlier stated problem – that of compensation for the build-up of the spray fan and other signal delays. We had to consider ways of making the most efficient use of the resolution provided by the CCTV camera based system.

Tests with spray guns, operated conventionally with electric solenoids outside the spray booth sending compressed air signals to open the spray gun, showed signal delays from 30 to 80ms and spray fan build-up time of around 50 to 100ms depending upon air pressure. Spray fan decay took even longer if the spray gun needle control cylinder had to exhaust itself back along the hose and through the electric solenoid. For this reason it was decided to use an intrinsically safe D.C. operated electric solenoid and mount it directly on the spray gun. This ensured that the maximum signal delay was then 20ms. The cylinder could then be operated from the atomising air line which was always a regulated line, thus ensuring reliable regular needle withdrawal times and also dispensing with the control air hose. Since the control cylinder was then operated by the clean dry atomising air, it was permissible to exhaust the control cylinder inside the spray booth via the intrinsically safe electric solenoid, thus achieving the shortest possible fan decay times.

In operation a reciprocating spray gun could be expected to travel at speeds up to 2m/s. This meant that a delay of 20ms became a distance travelled of 40mm and this represented two store locations displacement from the true position of the product and would have to be allowed for. The obvious method that came to mind was to devise a means of reading store ahead of the actual position of the spray gun. So a 7-bit adder circuit was inserted between the 7-bit line counter outputs and the memory address ports and a compensating value was added to the counter outputs. This compensating value was made adjustable by using a sixteen position switch, thus enabling a wide range of adjustment to be achieved. Two adjusting switches were provided, one for compensating up strokes and the other for down strokes.

These have been the major characteristics incorporated into this spray gun controller. There were other minor ones such as reverse compensation or mixed reverse and normal compensation, also individual spray guns spraying only on up strokes or only on down strokes.

Finally, there was one problem that occurred for which there was no electronic solution. On one installation using a vertical reciprocator with a 4.2m stroke length the optimum position for the camera was outside the factory in the car park looking through holes in two walls and the toilet block in order to see the product in the spray booth. Using a close-up lens only caused the light to concentrate in the centre of the lens with no discrimination at the edges. The solution adopted was to use a vertically positioned plane mirror to bend the line of sight of the camera through 70° so that with a standard lens the camera was then positioned alongside the spray booth wall looking along a line parallel to the conveyor and to the spray booth wall but outside the spray booth. This solution was then seen to yield other installation benefits and it became standard procedure to use a mirror in all subsequent installations.

FORGING: FEASIBLE ROBOTIC TECHNIQUES

R. Kelley, J. Birk, J. Dessimoz, H. Martins and R. Tella,
Robot Research Group, University of Rhode Island, USA

Presented at the 12th International Symposium on Industrial Robots, 9-11 June 1982, Paris, France

Abstract

Examination of the requirements for precision forging of airfoils for jet engines suggests that robot forging systems are feasible. Typically, a 950°C workpiece is being forged every 15 seconds. By employing robot systems which can endure conditions in the forge area better than humans, overall productivity gains can be expected. Furthermore, the operator could oversee several operations and be free to do jobs which are less repetitive and better suited to human capabilities.

An experimental robot system was assembled to simulate the operations discussed here. A total system approach showed that such forging systems are feasible. Experience with such systems will suggest further performance improvements.

Introduction

The forging process will be discussed in terms of the requirements for the precision forging of airfoils for jet engines. Other forging applications present similar technical challenges.[1,2,3,4] Typically, in the forging of airfoils, the workpieces are heated to a temperature of 950°C and produced at a rate of one forging operation every 15 seconds. Because of conditions in the forge area, the operator's endurance is an important factor in productivity. By employing robot systems in such environments, overall productivity gains could be expected. Moreover, the operator would no longer be required to stay in the midst of the oven-forge area. Rather the operator could oversee several operations and be free to do jobs which are less repetitive, better suited to human capabilities.

This paper describes feasible robotic techniques which overcome several impediments to employing robot systems for forging operations.

The forging process

The starting forms for the four operations in an airfoil forging sequence are shown in Fig. 1. The slug form feeds the extrusion operation, the extrusion form feeds the upset operation, the upset form feeds the block forge operation, and the block forge form feeds the final forge and trim operation. Typically, each forging operation consists of four steps:

☐ *Cleaning.* The cleaning step consists of tumbling the pieces to deburr and

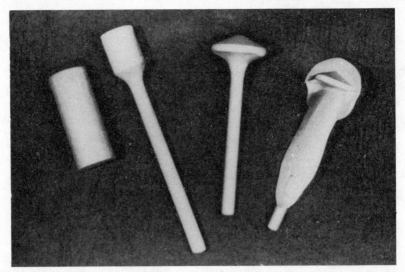

Fig. 1. Airfoil forging sequence (left to right): slug, extrusion, upset and block forge forms.

degrease the surfaces. Except for the slug form, the spent coating of lubricant from the preceding step is also removed. After tumbling, the pieces are stored in a basket.

☐ *Coating.* With the exception of the upset operation which uses only the graphite lubricant applied to the die, the other forms are also given a coating of lubricant. The pieces are coated away from the forge area by a worker other than the forge operator. They are taken from a basket one at a time, coated and set down to dry. The coating is ceramic, paint-like material which air dries in a few minutes to provide a fragile, easily scratched surface. When dry, the coated pieces are stacked carefully in a basket between layers of paper towels.

☐ *Heating.* Baskets of coated pieces are brought to the forge area. Prior to forging, they are heated to 950°C in a rotary oven. The forge operator takes a coated piece

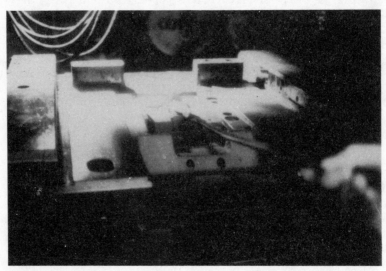

Fig. 2. Typical forge press loading operation.

Fig. 3. Typical trim press loading operation.

from the basket, grasping it with tongs. The cold piece is placed on the hearth of the rotary oven. A hot piece which has made a complete trip around the hearth is removed.

□ *Forging.* The forge operator takes the hot piece out of the oven using tongs. The hot piece is laid in the bottom forge die. This requires hand-eye coordination, particularly for the block forge form. This additional difficulty arises because the orientation of the airfoil varies from piece to piece. The stems are of variable length and at a variable angle to the airfoil. The forge operator holds the stem using tongs and adapts the loading motion to compensate for where the hot piece is relative to the die. A typical forge loading operation is shown in Fig. 2.

As the top die comes down and works the piece, tremendous forces are developed as the material is displaced. During the block forge and final forge operations, these forces cause the stem to move. To accommodate this motion, the

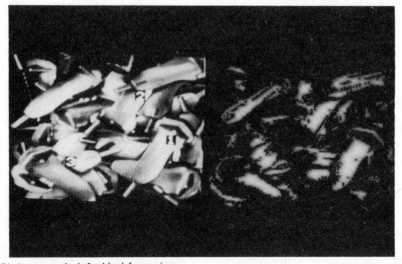

Fig. 4. Bin image analysis for block forge pieces.

Fig. 5. Special vacuum gripper

operator loosens the tongs slightly. As the top die goes up, the operator tightens the tongs and removes the piece from the forge die. Except for the final forge operation, the worked piece is dropped into a basket. For the final forging operation the worked piece is moved directly to the trim press and positioned in the trim die. The trimming placement is shown in Fig. 3. The operator deposits the trim scrap, held in the tongs, in a basket.

To assemble a robot system to implement forging operations as described above, the system must be able to perform the following activities:

○ Acquire a piece from a supply basket.
○ Orient a piece for coating (except upset).
○ Coat a piece with lubricant (except upset).
○ Place a piece in an oven.
○ Remove a piece from an oven.
○ Place a piece in a forging die.
○ Remove a piece from a forging die (for final forging).
○ Place a final forged airfoil in a trim press.
○ Remove the scrap from trimming.

Relevant research

The forging process is made up of activities which require the following skills:

● Bin picking: acquiring randomly oriented pieces from a supply basket.
● Pose estimation and pose refinement: estimating and improving the accuracy and precision of pose parameter values.
● Orientation manipulation: bringing a piece from an arbitrary position to a desired one with the proper pose.,

□ *Bin Picking.* There are many ways to remove pieces from a supply bin. Mechanical separators and feeders may be used. Although such feeders tend to be large and difficult to adjust, techniques for all forms except the block forge pieces are well known. Robot bin picking systems which use vision have been developed. Cylindrical slugs[5] and connecting rod castings[6] have been handled using a sensor equipped parallel jaw gripper. Electrical boxes[7] have been handled by a special vacuum gripper.[8] These systems are suitable for bin picking the forms encountered in the forging process as well.

☐ *Pose estimation and pose refinement.* When the degrees of freedom of the pose parameters can be constrained by resting a piece on a plane, the two translational and one rotational parameter values can be estimated by straight forward vision analysis techniques.[7] However, if the piece cannot be constrained, stereo methods must be employed.[9,10] For the forms typically encountered in the precision forging of airfoils, these approaches are effective.

☐ *Orientation manipulation.* The general problem of manipulation for reorientation involves the ability to regrasp the piece.[7] For this reason, the complete sequence of activities must be considered in the design of a robot forging sytem. It is desirable to minimise the requirements for regrasping a piece to achieve a particular pose.

Robot techniques

The robotic techniques which are used to implement a robot forging system are building blocks which form an integrated system solution to a piece handling problem. The key notions in constructing such a system are to segment the system along functional lines and to maintain piece pose knowledge from step to step. These notions result in minimising the number of times the piece is released by the robot gripper and in the use of the fixtures to maintain the pose of the piece when it is released. Functional segmentation divides the handling of uncoated pieces from coated ones. Hence a robot system architecture is suggested which uses one robot for the bin picking and coating operations and another for the oven and forge operations.

The robot which does the bin picking should be equipped with an appropriate gripper. For the first three forms (slug, extrusion, and upset) a parallel jaw gripper[1,2] employing simple binary vision holdsite selection[1,3] would be suitable. For the block forge form, the airfoil surfaces dominate the visual scene; hence, a vacuum gripper[4] would be suitable with a similar holdsite selection algorithm as for the other forms. The image analysis for block forge pieces is shown in Fig. 4. Three holdsites are labelled. A typical, special vacuum gripper is shown in Fig. 5.

Once a piece has been removed from the supply basket, the next activity is to coat it with a lubricant (except for the upset pieces). Since the coating can be easily damaged, coated pieces must be handled from the (stem) end. For the slug form, a simple V-groove inclined track with a stop is all that is necessary to orient the cylinders.[1] For extrusion and upset forms, simple sensing will determine which is the thicker end; then placement in a similar track would suffice to orient these forms for further handling.

For the block forge form, however, the special vacuum gripper is required. It provides a support plane which constrains the pose estimation problem in a manner similar to placing the piece on a table. The binary images used to determine the translation, rotation, polarity and state are shown in Figs. 6 and 7. The rectangular area tests locate the narrow stem end. The profile view locates the position of the neck on the stem (square box) as well as locating nub (state identification). This information is used to permit the airfoil to be hung in a slotted coating fixture.

The coating fixture slot constrains the location of the end of the stems of the coated pieces. Since this part of the piece is not worked, the robot can grasp the piece by the stem for transferring the coated piece to the hearth of the rotary oven.

To constrain the pieces in the oven and to minimise any damage to the coating

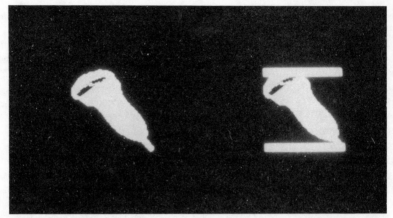

Fig. 6. Pose estimation analysis, face view.

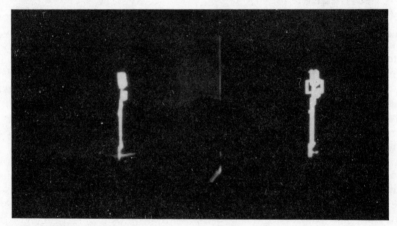

Fig. 7. Pose estimation analysis, profile view.

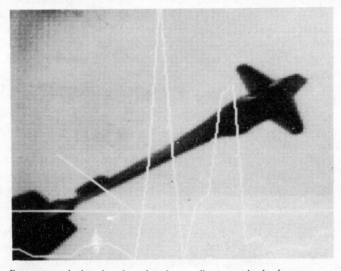

Fig. 8. Pose refinement analysis, edge view showing gradient magnitude along scan.

Fig. 9. Pose refinement analysis full face view.

due to vibrations and the like, special oven fixtures can be employed. This way, transfers from the coating fixture are fixed point-to-point motions. On removal, the robot acquires hot pieces by executing a fixed sequence of pickup actions.

Pieces removed from oven fixtures are highly constrained. At most, then, pose refinement methods might be needed before the piece can be put into the forge die. Because the pieces are glowing, vision analysis must be rapid (to prevent cooling) and capable of dealing with the high level of infrared radiation. Since robot manipulations are most easily taught using cold, coated pieces, the use of filters to block the IR and the addition of external lighting to illuminate the piece is recommended. If subtle features are to be detected, grey scale image analysis which is restricted to extracting the minimum amount of information is feasible.

In the case of the block forge pieces, grey scale images allow a coordinate system to be constructed through the spatial locations of features on the piece. Shown in

Fig. 10. Template for dovetail structure.

Figs. 8 and 9 are edge and full face views of a block forged piece. Template matching allows the dovetail structure (Fig. 10) and the nub structure (Fig. 11) to be located in space. This information allows the corrections to be computed which enable the robot to place the piece in the bottom forge die.

When the extrusion or block forge forms are worked, the robot gripper can maintain its grasp on the stem. To isolate the robot arm from the forces generated by the forging operation, an accommodating mechanism is required. For the block forging of the extrusion, the main concern is the removal of the forged piece from the die; any way of relaxing the grasp on the stem would be acceptable. However, for the final forging operation, it is necessary also to perform a trimming operation while the piece is still hot. Since the piece is accurately located in the forging die, the transfer from there to the trim die always follows the same trajectory.

Because the robot is holding onto a moving stem, a way to mimic the action of a human operator is desired. One solution is to use an accommodation linkage, A/link, such as shown in Fig. 12. The A/link is maintained in a standard configuration until the top die descends. Then it momentarily relaxes to permit the gripper held stem to relocate. The A/link temporarily assumes some accommodating configuration while allowing the robot arm to maintain its pose. The A/link provides the mechanical connection between the robot arm and the piece in the forge die. The necessary arm joint motions to bring the piece from the forge to the trim die are easily computed.

Summary

The precision forging of airfoils for jet engines can be performed by robot systems. The techniques which are needed for a robotic solution have been identified. The details of these techniques have been presented elsewhere. These techniques were organised to address the issues of forging. To meet the special requirements of specific applications, additional engineering problems will arise and have to be solved. The workplace should also be equipped with sensors and other devices as good engineering practice demands.[11] What is desirable for hard automation is usually necessary for robot systems.

An experimental robot system was assembled to simulate the operations

Fig. 11. Template for nub structure.

Fig. 12. A / link accommodating linkage.

described here. A total system approach to these activities showed that such forging systems are feasible. Experience with such systems will suggest further performance improvements.

Acknowledgement

The experimental robot systems and the techniques referred to in this presentation came about through the efforts of the members of the Robot Research Group of the University of Rhode Island, past and present, especially A. Vinci, J. Crouch and J. Hall.

References

[1] I. Franchetti, P. Vicentini and L. deTogni, 'Automising a work place in forging by robot'. *Proc. Sixth International Symposium on Industrial Robots,* Nottingham, UK, 24-26 March 1976.

[2] E. Appleton, W. B. Heginbotham and M. Kohno, 'Design study and feasibility trial for a robot blacksmith'. *Proc. Eighth International Symposium on Industrial Robots,* Stuttgart, W. Germany, 30 May-1 June 1978, pp. 528-543.

[3] E. Appleton, W. B. Heginbotham and M. Kohno, 'Experimental study into the use of an industrial robot as a manipulator for open die forging of light work pieces'. *Proc. Ninth International Symposium on Industrial Robots,* Washington DC, USA, 13-15 March 1979, pp. 709-728.

[4] A. K. Sengupta, E. Appleton and W. B. Heginbotham, 'Ring forging with an industrial robot'. *Proc. Tenth International Symposium on Industrial Robots,* Milan, Italy, 5-7 March 1980, pp. 29-42.

[5] R. Kelley, J. Birk, H. Martins and R. Tella, 'A robot system which acquires cylindrical workpieces from bins'. *IEEE Trans. Systems, Man and Cybernetics,* to appear in Vol. SMC no. 12 (2), pp. 204-213 (March/April 1982).

[6] R. Kelley, J. Birk, J. Dessimoz, H. Martins and R. Tella, 'Acquiring connecting rod castings using a robot with vision and sensors'. *Proc. First International Conference on Robot Vision and Sensory Controls,* Stratford-upon-Avon, UK, 1-3 April 1981, pp. 169-178.

[7] J. Birk, R. Kelley and H. Martins, 'An orienting robot for feeding workpieces stored in bins', *IEEE Trans. Systems, Man and Cybernetics,* Vol. SMC-11 no. 2, pp. 151-160 (Feb. 1981).

[8] J. Birk, R. Kelley and R. Tella, 'Apparatus for acquiring workpieces from a storage bin or the like', US Patent No. 4,266,905, 12 May 1981.

[9] N. Chen, J. Birk and R. Kelley, 'Estimating workpiece pose using the feature points method', *IEEE Trans. on Automatic Control,* Vol. AC-25 no. 6, pp. 1027-1041 (December 1980).

[10] H. Martins, J. Birk and R. Kelley, 'Camera models based on data from two calibration planes', *Computer Graphics and Image Processing,* Vol. 17, no. 3, pp. 173-180 (March 1981).

[11] J. Saladino, 'Upset forging with industrial robots'. Technical paper MS80-704, Robotics International of Society Manufacturing Engineers, Dearborn, MI, 1980.

AUTOMATIC CHOCOLATE DECORATION BY ROBOT VISION

A. J. Cronshaw, Patscentre International, UK

Presented at the 12th International Symposium on Industrial Robots, Paris, 9-11 June, 1982

Abstract

A prototype machine has been developed for the decoration of chocolates with piped patterns. The chocolates are decorated whilst on a conveyor belt by a robot arm equipped with a piping nozzle. The robot is guided by a machine vision system which enables the robot to adapt intelligently to the exact position and orientation of the chocolate. The nozzle is therefore directed to apply a piped pattern of chocolate mixture to the top of the chocolates with correct alignment.

The system is also capable of recognising mis-shaped chocolates such as doubles (two stuck together) or over-size or under-size chocolates. The paper includes a description of the software techniques for visual recognition and guidance of the robot. The cost and performance of the system are also discussed.

Introduction

The decoration of chocolates – and in fact the handling of chocolates generally – is largely a manual operation in many confectionery companies. Operations such as chocolate decoration and chocolate packing are often very labour intensive. The continual quest for a higher quality product in the face of increasing labour costs and decreasing robot cost has spurred the development of automated inspection techniques.

Automation of these tasks needs to be flexible for all but the largest manu-facturers: batch production is often involved, and product changes are often needed to meet changing market demands.

Patscentre International is actively involved in this field through the develop-ment of versatile robots equipped with sophisticated image processing systems. Patscentre's robot systems can be reprogrammed to meet:

○ change-overs from batch to batch
○ changing product lines

Furthermore the 'intelligence' and adaptability of 'seeing' robots means that these systems can be introduced into existing factories with minimum peripheral implications. It is not necessary to invest in feeders of magazining systems for example. A seeing robot can often work directly from existing conveyors and factory storage systems.

The very nature of confectionery products is also a good reason for this approach. Chocolates are fragile and easily scratched (and the lady does not like scratched

T.V. camera

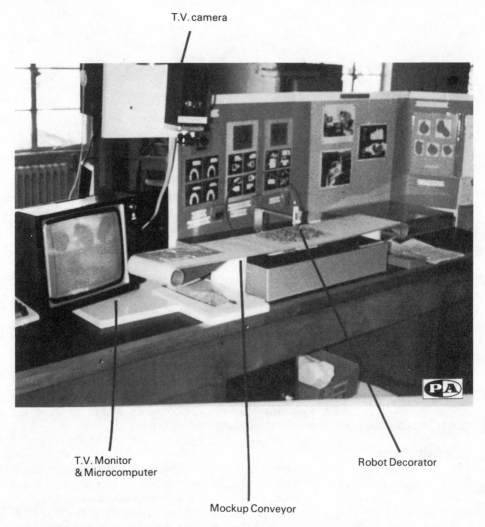

T.V. Monitor
& Microcomputer

Robot Decorator

Mockup Conveyor

Fig. 1. Robot Chocolate Decorator System.

chocolates!) A seeing robot eliminates the need to feed chocolates to the robot by mechanical feeders which can damage the product. This can lead to a variable size product which is difficult to handle except by an intelligent and adaptive system.

Existing manual methods

Decoration can most conveniently be carried out at one of two stages in the manufacturing process. The first opportunity is whilst the chocolates are riding along the enrobing conveyors. Here the chocolates are well regimented in regular rows and columns.

Variations in position and orientation are however inevitable: as the chocolates change over from one belt to another slight disturbances can result. The exact position and orientation of the chocolates is not a repeatable absolute. To apply a decoration the only methods to date have been to apply non critical patterns such as

Fig. 2. Robot Decorator Arm.

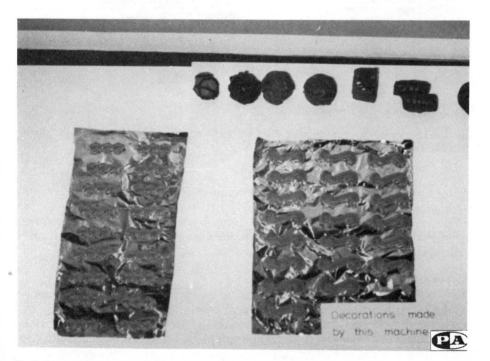

Fig. 3. Decorations made on the Prototype Machine.

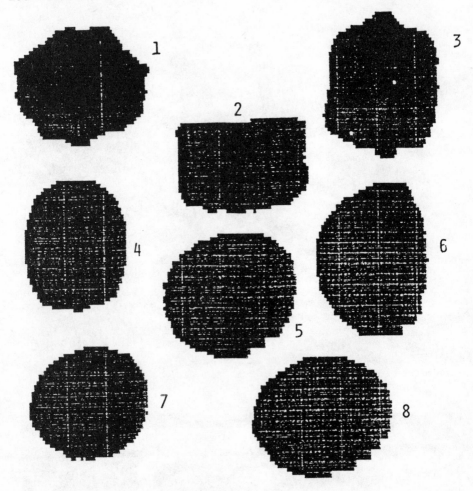

Fig. 4. Digitised image of eight chocolates.

Key: 1 Apricot Parfait, 2 Hazelnut slice, 3 Alpini, 4 Mandarin Truffle, 5 Continental Cup, 6 Rum Marzipan, 7 Coffee Truffle, 8 Walnut Truffle.

stripes automatically, or if a piped decoration is required to employ manual methods.

The second opportunity to apply a decoration is when the chocolates have been transferred to factory storage trays. Here the chocolates are randomly arranged. Manual methods alone have been used to date.

Robot decorator

The prototype Robot Decorator system is shown in Fig. 1. A TV camera is mounted overhead looking down vertically at chocolates on a conveyor. A microcomputer (a DEC LSI-11) is used to analyse the visual data from the camera and to control the robot arm (Fig. 2). The arm has two degrees of freedom in X and Y coordinate directions. The prototype system uses ambient lighting though obviously more robust methods would be used in a production system.

The system operates by scanning chocolates at one point on the belt and then a short distance downstream the chocolates are decorated. There is a certain amount of queueing of data to accommodate this gap. This scheme means that the decorating head never appears in the imaging area.

Recognition software

The TV picture is digitised onto a 256 × 256 matrix of binary pixels and stored in the LSI-11. Feature extraction routines are applied to this data to extract key parameters about chocolate shapes. These include:

○ Area,
○ Perimeter,
○ Centre of area,
○ Radii signature,
○ Corner extraction.

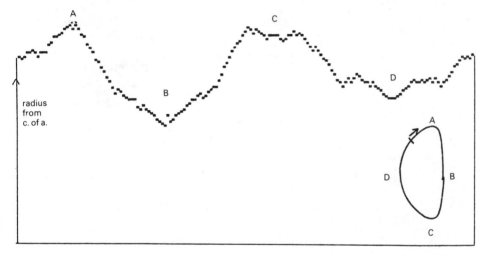

Fig. 5. Signature of routine RADII for Rum Marzipan.

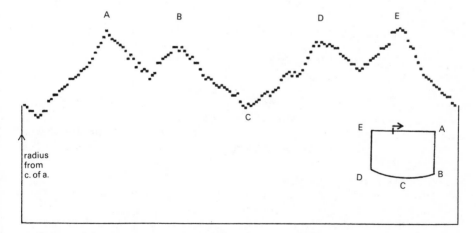

Fig. 6. Signature of routine RADII for Hazelnut Slice.

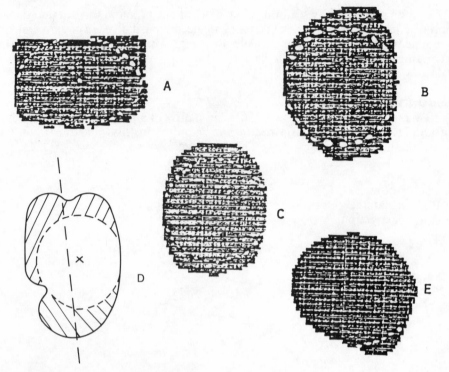

Fig. 7. Extraction of 'Prominence' Feature.

The radii signature is the result of measuring the distance from the centre of area each of the points on the perimeter. The signatures for two of the chocolates digitised in Fig. 4 are shown in Fig. 5 and Fig. 6. Peaks in the radii signature correspond to corners and valleys to sides.

Prominence

A corner extraction routine has been developed to handle the rounded 'corners' that are typical of chocolate shapes. The feature has been named 'prominence' to distinguish it from the usual sort of corner.

Prominence is defined a shown by the shaded area in Fig. 8. Mathematically it is the region enclosed by the smooth spiral joining R_A and R_B (both local minima) and the edge itself. Examples of this feature are shown in Fig. 7.

The result of extracting the prominence feature is to isolate the rounded corners of a given shape and to characterise it by the size and position of the corners. The size of the area is a key factor in gauging the extent of the corner. And the centre of area of the prominent areas is a robust indication of where the corner is located. For chocolates with round corners this routine is a key technique to find the long axis (i.e. orientation) reliably despite the presence of variations in the perimeter shape.

Rejection of mis-shapes

The prototype system has been configured to recognise three types of chocolate as shown in Fig. 2 and 3. The area of chocolates is used as a simple discriminator. Chocolates that exceed a maximum threshold are rejected by the system. Similarly chocolates that are too small are rejected and an error message is logged.

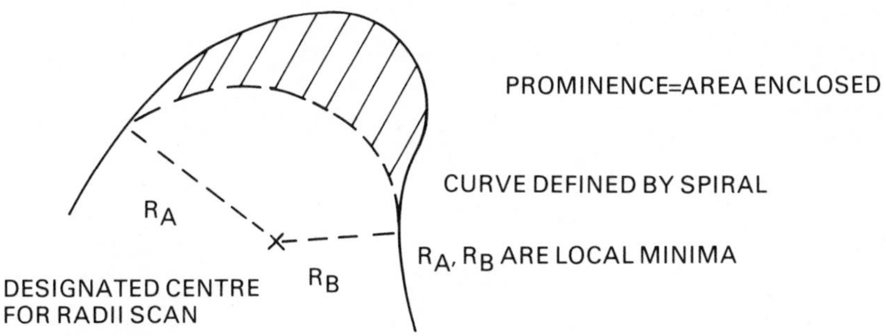

Fig. 8. Definition of 'Prominence'.

A more detailed study of the question of recognition of mis-shapes has been published in connection with automatic chocolate packing.[1]

Performance

The performace of the system is limited in the prototype by the speed of the decorating arm. The capacity of the microcomputer to recognise shapes has been found to be around 250mS/chocolate. The decorating arm is able to utilise only 17% of this capacity with a cycle time of around 1.5 seconds. It is suggested that a production version machine would use a number of arms per scanning system.

The cycle time of the software has been optimised by writing the software in a medium level programming language.[2] 'EXTRA' offers the speed of program execution of assembly language together with the high-level features of a high-level language.

A feature of the EXTRA language is the extensive use of machine orientated objects such as bits bytes and words. The EXTRA compiler is able to generate one line machine codes for most operations on these quantities.

References

[1] Cronshaw, A. J. 'Software development for the visual recognition of chocolates' *Proc. 2nd Int. Conf. on Assembly Automation,* Brighton, IFS (Publications) (May 1981).

[2] Cronshaw, A. J., Heginbotham, W. B. and Pugh, A. 'Software techniques for an optically tooled bowl feeder'. Conf. Trends in On-line Computer Control, Institution of Electrical Engineers, London (March, 1979).

Chapter 7
COMMERCIAL ROBOT VISION
SYSTEMS

There are a number of commercial vision systems on the market. The papers describe five of the most important systems that are available now.

OMS – VISION SYSTEM

P. F. Hewkin and H.-J. Fuchs
BBC Brown, Boveri & Cie AG, West Germany

Abstract

OMS is a commercial device for the sensing and high-speed analysis of optical data. It may be programmed by the user to perform optical processing tasks, such as the recognition of objects and the measurement of characteristic quantities, and to communicate with other devices. 'Interactive', 'Master' and 'Slave' modes of operation are possible. Typical operation times are under one second and applications include robot vision, sorting, handling and quality control problems.

Introduction

Recent developments of micro-processor technology have made feasible the concept of a general-purpose optical sensor. This differs from the previous generation of dedicated sensors in that it is flexible and not developed especially to perform a single task.

This paper refers to OMS (Fig. 1)[1,2,4] which has been designed for use in a wide range of applications. OMS is well established as a production device and practical experience in laboratory and industry,[1,6,7,8] has been gained and is being exploited in continuing development.

System features

These may be divided into two classes, optical processing features and peripheral features. OMS features have been chosen to cover a broad spectrum of optical analysis tasks and to suit various modes of application.

The basic OMS optical processing features are briefly listed in Table 1. OMS may be specified by the purchaser to include some or all of these features. Existing systems can be up-graded by the addition of slide-in modules (Fig. 1). Picture data is analysed in binary form and the assumption is made that a suitably, high-contrast picture (in either transmitted or reflected light) can be obtained by using tube, matrix or line-scan cameras. The accuracy of measurements is limited primarily by area represented by a pixel (picture size 256 × 415 pixel for raster cameras, 1024 × 512 for line-scan-cameras). Calculations are carried out in sub-pixel units to avoid rounding problems. A detailed discussion of optical features is contained in [2].

The mode of application of OMS sets user requirements additional to those of optical processing. Such requirements vary from application to application. Three basic modes may be defined (Fig. 2).

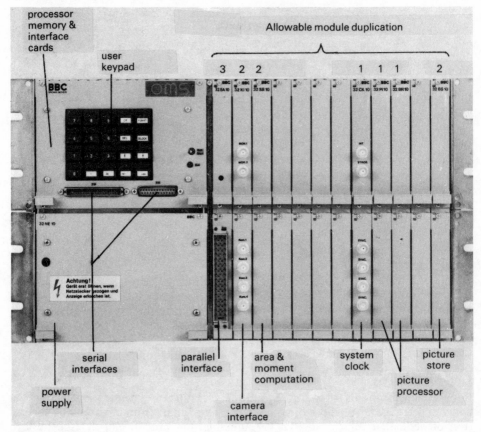

Fig. 1. OMS (front view).

1.   Stand-alone. OMS is used interactively by a human operator who observes the selected field of view on a TV monitor and enters appropriate commands through a keyboard. Typical applications are remote object recognition and measurement (quality control) and OMS programme development.

2.   Master. With the aid of data won from optical analysis, OMS is used to drive a slave device such as a sorting unit for a conveyor system. The operation is pre-programmed in OMS by the user. OMS issues instructions to the slave to implement actions such as 'STOP', 'REJECT OBJECT', 'SELECT OBJECT TYPE X'.

3.   Slave. OMS is used to provide an analysis of optical data to a master computer which controls an operation. Data such as object type and the co-ordinates of a desired point can be provided upon request. Typical applications include automated assembly.

Table 2 outlines some of the basic peripheral features of OMS and indicates their relevance to each of the three modes.

Programming

The system may be programmed via the 20 key OMS front panel, via an optional ASCII keyboard or from a host computer. OMS instructions are selected by the user from a series of system functions (Table 3). Each function involves a simple

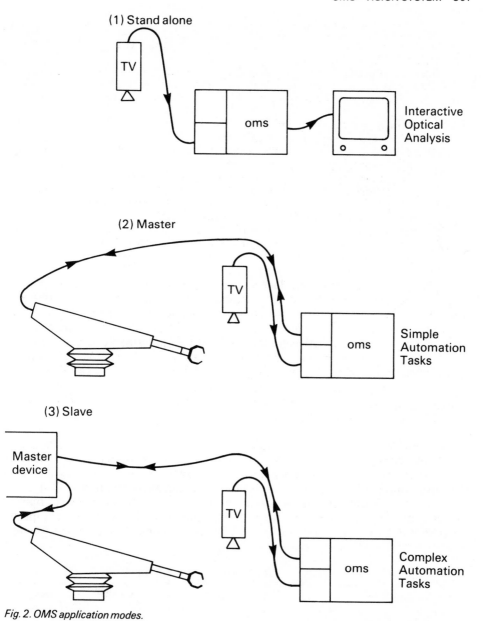

Fig. 2. OMS application modes.

question and answer procedure with the current question and user's response being superimposed onto a TV monitor showing the selected field of view if required. An invalid input results in the display of an error code. The error may be corrected by entering a valid input without restarting the function from scratch. Alternatively, errors may be reported to an external device which takes appropriate action (e.g. by sending a new command to OMS).

In addition, the user may create his own functions by entering a learn phase and going once through the various commands which are required (teaching by

Table 1. OMS Optical Processing Features

1)	Selection of camera input channel and frame containing the desired region for analysis.
2)	$2 \times 512 \times 512$ point binary picture stores.
3)	Selection of display mode, including analogue/binary, positive/negative, AND and EXOR of two camera inputs or of camera input and picture store.
4)	Fast computation of picture processing operations. Bit slice processor with 125ns circle-time.
5)	Area, Centre-of-Area coordinates available in real time.
6)	Translation of measurements into real units.
7)	Object learning and position independent recognition and classification. User-interrogation of recognition criteria allowed.
8)	Object position measurement (x, y, ϕ) relative to learned position.
9)	Tracking of learned point (not necessarily centre of area) on, or relative to, an object.
10)	Separation of multiple objects – up to 256 objects in one picture. Sorting of such objects according to size or in rows or columns. Individual analysis.
11)	Run-length analysis of vectors positioned relative to recognised objects, e.g. space interrogation for robot gripper. Analysis of corners or local features.

showing) and then storing the learned 'macro' function under a user specified name to be recalled by a simple command. Within this facility the user may program infinite or finite program-loops, various forms of conditional wait and also data 'handshakes' to various interfaces. Calculator-like features are available to facilitate data transfer between functions and a programmable logic feature may be used, for example, to make the program-path dependent upon the results of optical processing. Once programmed, the user functions and all variables which contribute to the programmed operating mode can be dumped onto cassette or paper tape. User data can also be held in non-volatile memory so that OMS automatically recovers after power failures.

Interfacing
OMS may be configured to accept up to eight inputs from raster TV cameras or up to four line-scan cameras. Two outputs are available for monitors and synchronisation signals for cameras (and or monitor) and also generated by OMS.

Data interfaces to other devices (e.g. terminal or master/slave device) may be selected by the purchaser and programmed interactively by the user. Options include standard DEC serial interfaces (V24, 20mA current loop etc.) as well as a

Table 2. Requirements for OMS

Mode of Application Feature	Stand-alone	Master	Slave
Sensible interactive display	†		
Simple programming language	†	*	*
User-composed programmes	†	†	*
Flexible, programmable interfaces		†	†
Robot support software		†	†
Calculator-like memory and simple arithmetic	*	†	
Internal timing and conditional waits		†	*
Ability to store and recall user programmes	†	†	*

† = Most desirable.   * = helpful.

Table 3. OMS System Functions

No.	Description	Comment
1	Display of available functions	Both system and user functions
2	System initialisation, loading and dumping of user programmes	OMS user programmes can be dumped on tape or cassette
3	Interrogation of programme/data addresses	For maintenance or debugging
4	User-programming	For the creation of user functions
5	Input/output specification	Selection of interface and programming of handshakes
6	Programmed loop	
7	Programmable conditional wait	For process synchronisation
8	Programmable delay	
9	System calibration	Requires test picture to correct for camera distortion and to learn calibration units
10	Scaling between pixels and calibration units	
11	Camera, mode and field of view selection	
12	Automatic setting of binary picture threshold	For minimum sensitivity to illumination fluctuations
13	Parameter specification for learning and recognition	User specified or automatic
14	Object learning	User specified name and class
15	Learning of orientation and reference points	
16	Object recognition	
17	Measurement of orientation and translation	Relative to learned position or absolute. Enables tracking of learned points
18	Measurement of area and x, y position. Comparison with picture store	
19	Programmable logic	For conditional programme branch etc.
20	Separation and analysis of multiple objects	
21	Calculator logic	$+- \div \times$ operations on programme data. Also 10 independently addressable user memory cells
*22	Vector analysis	Analysis of vectors drawn in picture store. Absolute or relative to recognised objects
*23	Picture store management	Loading, Display and Analysis of the contents of the two OMS 512×512 pixel binary picture stores
*24	Picture rotation	Rotation of user-specified section of picture store by desired angle
*25	Memory	Extended storage for user-data
*26	T3	Support for Cincinnati-Milacron T3/HT3 robots. (Reference 5, 8)

*Special functions not included in standard OMS software

programmable parallel interface with 16 input and 16 output channels. Special interface software has been designed to support several industrial robots (e.g. T3 from Cincinnati Milacron and PUMA from Unimation[6,8]). Serial interfaces are buffered by OMS and features such as auto-linefeed and echo (for terminals) can be selected interactively. Handshakes using almost the full ASCII character set

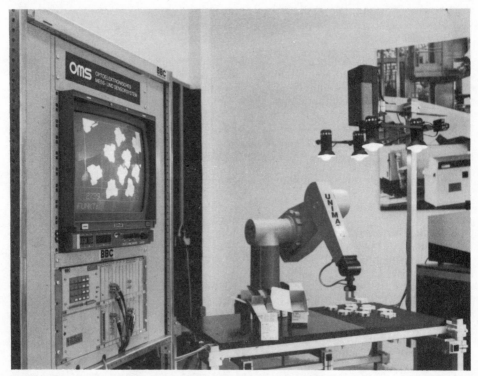

Fig. 3. OMS – PUMA application – Shows a PUMA robot packing randomly orientated circuit breakers into cartons under the supervision of an OMS vision system. OMS emulates the robot's programming terminal, sending appropriate commands to the VAL programme when the robot desires the coordinates of the next object to be packed. Standard OMS and PUMA systems were used without additional hardware.

(including control characters) may be preprogrammed to enable the transfer of data via the selected interface. This allows the transmission of intelligent text as well as data e.g. object No. 7, position x = 57mm, y = 29mm, phi = 3501 1/10 deg, to a human operator as well as coded instructions to another device. With the aid of the programmable logic feature, the parallel interface may be used for complex sorting or processing operations, for example, in conjunction with hardwired control systems. Additional interfacing possibilities are offered by programmable waits for conditions such as 'suitable area in field of view' or 'slave device ready' or the receipt of a desired ASCII string.

Operation timing

A basic requirement of an optical sensor is that it can operate at a speed compatible with the other elements in the system (assembly robots, conveyor belts etc.). Cycle times for OMS are dependent upon the operations performed and, in some cases, upon the complexity of the field of view to be analysed. OMS employs separate processors, for program management and for picture processing, which run simultaneously during many operations. Table 4 presents a list of typical operation times for OMS and the factors which affect them. Typical cycle times for optical processing and decision making by OMS lie under one second. This makes OMS suitable for a large number of practical applications.

Table 4. OMS Operating Timing

Operation	Typical Time Requirement	Comment
Computation and output of white area in field of view	25ms	Object independent
(as above) with additional output of centre of area co-ordinates	60ms	Object independent
Object recognition with output of object number, quality of recognition, 'next-best' object number and its recognition quality	70ms-190ms	Dependent an object geometry, desired recognition parameters and number of learned object
(as above) with output of object orientation and co-ordinates of learned point	100ms-280ms	Dependent on desired accuracy and on object recognition conditions
Separation of multiple objects	300ms-2000ms	Dependent upon picture complexity
Recall of individual objects	10ms-20ms	Dependent upon picture complexity
Tracing and interpretation of a pre-programmed vector in the picture store	200ms-500ms	Dependent upon number of colour charges in the vector
Rotation of selected area of picture store about desired axis	500ms-2000ms	Dependent upon picture complexity

Applications

OMS is being used in research and industry throughout Europe. Industrial applications include unloading pallets, sorting of metal fittings, measuring the position of workpieces and containers, programme selection for paint spraying apparatus, quality control and handling of small parts.

The system has been used together with robots from Unimation[6], see Fig. 3, Cincinnati-Milacron[5,8] and DEA.[7] The OMS interface support software makes communication with a large number of other robots easy for the OMS user to arrange to his own specifications.

Conclusion

The OMS vision system is not only capable of collecting and analysing optical data, but also of using this data to provide intelligent communication with other devices (in particular, with industrial robots). The flexibility of its modular structure and the ease of programming make it suitable for a wide range of applications.

References

[1] R. Karg and O. E. Lanz, 'Experimental Results with a Versatile Optoelectronic Sensor in Industrial Applications'. *Proc. 9th Int. Symp. on Industrial Robots*, Washington DC, USA (Society Manufacturing Engineers) (March 1979).
[2] P. F. Hewkin and H.-J. Fuchs, 'New features of the OMS Vision System'. 2nd Int. Conf. on Robot Vision and Sensory Control, Stuttgart, Germany (2-4 November 1982).

[3] H.-J. Fuchs and P. F. Hewkin, 'Roboteranwendung mit dem Optoelektronischen Sensor OMS'. Fachtagung 'Industrieroboter Chancen und Probleme für unsere Betriebe, Universität Linz, Austria (28-30 September 1982).

[4] H. Holsmölle, 'Sehende Roboter in der Handhabetchnik'. BBC Heidelberg (1981).

[5] 'External Function RS-232 (Option) with restructured Software'. Cincinnati-Milacron Co., Cincinnati, USA (1980).

[6] P. F. Hewkin and H.-J. Fuchs, 'Erfahrungen mit dem Robotersichtsystem OMS/PUMA. BBC Heidelberg (1981).

[7] I. Schmidtt, 'Montageaufgaben für sensorbestückte Industrieroboter'. Presented at conference 'Erfahrungen mit Industrierobotern', Munich, Germany (28-31 October 1981).

[8] P. F. Hewkin and H.-J. Fuchs, 'OMS/T3 Robot Vision System'. BBC Heidelberg (1981).

THE PUMA/VS-100 ROBOT VISION SYSTEM

Brian Carlisle and Scot Roth, Unimation Inc., USA and Jerry Gleason and Dennis McGhie, Machine Intelligence Corp., USA

Presented at 1st International Conference on Robot Vision and Sensory Controls, Stratford-upon-Avon, UK, April 1-3 1981

Abstract

A third-generation robot system with integrated visual sensing is described. It is composed of Unimation's PUMA robot and Machine Intelligence Corporation's VS-100 vision system, and is controlled by Unimation's VAL language. With the ability to determine the position and orientation of objects, this advanced robot system can significantly reduce the cost associated with precise indexing. Technical details of the PUMA robot, the vision system, and the interface are described. Calibration procedures and the vision-to-robot transformation are discussed. The necessity for robot accuracy is explained and overall system accuracy is specified. A brief example program is included.

Introduction – The Univision system

The Univision system is a complete robot/vision system comprised of a Unimation Inc. PUMA series manipulator and a Machine Intelligence Corp. VS-100 vision system. The system is trained and operated through VAL, Unimation's easy-to-use language. With the ability to determine the position and orientation of objects, this advanced system can significantly reduce the cost associated with precise indexing. In addition, the visual algorithms employed allow in-process inspection of work-pieces in a material-handling or assembly system. The vision system is 'trained by showing' i.e., simply by taking pictures of sample parts. The robot is programmed using VAL, a high-level robot control language consisting of English words such as MOVE, DEPART, and LOCATE. The system can be set up and calibrated in a few minutes. The vision system can process image data while the robot is moving, allowing typical cycle times of three-five seconds for a part acquisition and transfer.

PUMA Series manipulators

There are now three machines in the PUMA series of Unimation Inc. manipulators. The Model 250 is a six axis machine with a 1kg capacity, a reach of 0.4m, and tip speeds up to 3m/s. The model 500 and model 600 are five and six axis machines, respectively, with a 3kg capacity, a reach of 0.86m, and tip speeds of 1m/s. PUMA series manipulators are driven by DC servo motors with incremental optical encoders. Major axes have fail-safe brakes. Each joint has a two-stage, zero backlash gear train. The upper and lower link designs are of a unique monocoque construction which maximises the strength-to-inertia ratio.

Each PUMA robot has its own dedicated computer controller. The heart of the system is the DEC LSI-11 computer. User programs are stored in RAM and the VAL software in EPROM. The servo system consists of a microprocessor and power amplifier for each joint. Each microprocessor receives a joint angle command from the LSI-11 and closes a software servo loop. This architecture allows dynamic updates of servo constants.

The entire controller including a regulated DC power supply is self-contained in a 19 inch rack mountable enclosure. All peripherals plug into front panel connectors.

The teach pendant is microprocessor based and consists of joint drive switches, mode selection switches, a tool clamp switch, a teach speed potentiometer and an alpha-numeric display. The display is used for error messages as well as many helpful status messages.

VAL language

The PUMA system operates with a high level language called VAL. In addition to being a sophisticated programming language developed for assembly, VAL is a complete robot control system.

VAL consists of a full set of English language instructions for robot teaching and program editing. The VAL control system has been designed, in addition to operating the PUMA, to readily communicate with other computer-based systems

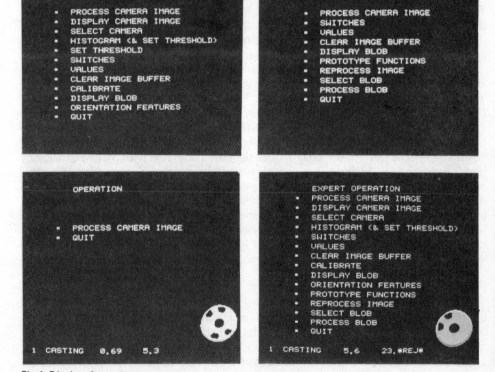

Fig. 1. Display of sample menus.

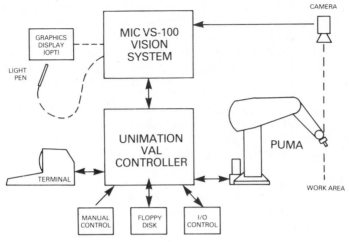

Fig. 2. Univision system.

such as vision and force sensors. The VAL vision interface instructions will be described in detail below.

MIC VS-100 vision system

The MIC VS-100 vision system locates visible objects and communicates their positions to the robot for manipulation. The vision system initially thresholds a video grey-level image into a binary image and thereafter works with blobs, i.e., object silhouettes. The location of an object that the vision system provides is simply the 2-D centroid of the blob. The orientation of an object is computed based on user selectable criteria such as angle of largest radius or angle towards largest hole.

The overall success of the system depends on object recognition. The vision system characterises blobs on the basis of such features as area, perimeter, minimum and maximum radii, and number of holes. To recognise objects, the vision system is 'trained by showing' sample objects to the system. The user simply defines a prototype object by name and then shows it to the system in different locations, while the system accumulates statistics on the object's visual features.

The key attributes of the VS-100 machine vision system can be summarised in four major areas:

□ First, to simplify user interaction, a light-pen-controlled menu was developed which allows a relatively unskilled person to set up and operate the system. There are four major menu sections which consist of SYSTEM SET UP, TRAINING, OPERATION, and EXPERT OPERATION. 'SYSTEM SET UP' is used to select the camera, set the field of view, lens, aperture, focus, and threshold level, and perform spatial calibration. 'TRAINING' is used to program the VS-100 (by showing) to recognise parts in each of their stable states. Selection and optimisation recognition features can be accomplished using an automatic discrimination matrix. 'OPERATION' allows activation of the VS-100 using the training data and control parameters previously obtained. 'EXPERT OPERATION' has several additional functions that allow a user to optimise the VS-100 for a particular application. Timing information is available to analyse the trade-off between speed and reliability in a given application. (See Fig. 1 for sample menus).

☐ Second, the VS-100 allows direct connection to stable, high-resolution, solid-state cameras such as the G.E. TN2500 (240 × 240 array) or the Reticon LC600C256-1 (256 × 1 array) for moving-line applications.

☐ Third, a complete computer communication interface has been developed to allow remote control of the VS-100 from a higher level or process-control computer. This interface essentially allows a user's computer to treat the VS-100 as a peripheral device that processes visual information. An example of this capability has been demonstrated with the Unimation PUMA robots. The hardware portion of the interface consists of a standard DEC 16-bit bi-directional TTL parallel interface module. The software communication protocol was especially developed to provide reliable high-speed inter-computer communication.

☐ Fourth, the hardware in the VS-100 was designed for a rugged factory environment. A conservatively rated high-quality switching power supply provides excellent line voltage and noise isolation for reliable operation. A battery back-up unit is available for more severe power line environments.

The Univision system

The Univision system combines the PUMA series robots with the VS-100 vision system. This powerful new system has the ability to acquire randomly located workpieces and hence reduces jigging, fixturing, and part presentation requirements. The ability to acquire parts from a moving conveyor belt is being developed. This adds tremendous flexibility to materials handling systems. The vision system can process image data while the robot is moving, allowing typical cycle times of three-five seconds for acquisition and transfer of a part.

The robot-vision interface

In the Univision system (Fig. 2), the vision system is slaved to the VAL robot control language. As a consequence, a number of vision related instructions have been added to the language. These instructions permit the user to calibrate the system, train vision prototypes, identify and locate workpieces, and store or load vision data from the Unimation floppy disc unit. During operation the robot can command that a picture be taken, and subsequently ask the vision system to locate a particular part. The vision system will process the blobs in the picture, and attempt to match them to prototypes. Among those blobs which match the prototype, the best match is chosen. If a match is found VAL defines a transformation with the same name as the requested part which represents the location and orientation of the object relative to the camera reference frame, as determined by the vision system. If no match is found VAL can branch to another program step for remedial action.

Calibration

In order for the robot to acquire parts located by the vision system, there must exist a common cartesian frame of reference. This is accomplished during system calibration. First, the vision system is calibrated to establish an accurate two dimensional coordinate system in the field of view. Next, the camera coordinate system must be related to the coordinate system of the robot. This is accomplished by determining a camera-to-robot coordinate transformation. A coordinate transformation is an operator which describes translation and rotation of a coordinate system from one location to another. Thus the camera-to-robot transformation describes the translation and rotation required to bring the camera coordinate system into coincidence with the robot coordinate system.

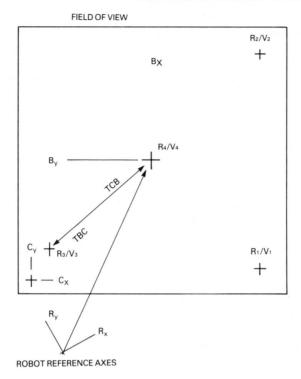

Fig. 3. Robot/camera transformation.

Vision system calibration

The vision system must have its coordinate system scaled in the same units as the robot coordinate system. The vision system is calibrated using the robot to measure a distance in the field of view. This is accomplished by the following method: A 'pointer' tool is held in the robot hand. A flat disc with a centre hole is placed in the field of view. The robot is moved to place the pointer in the centre hole of the disc. The robot location is recorded by typing HERE R1 on the terminal. The robot is moved out of the field of view without disturbing the disc. The commands PICTURE and FINDHEAP V1 are typed on the terminal, causing the vision system to determine the location of the disc in the field of view. The procedure is repeated for a second location (using R2 and V2) yielding two robot locations and two corresponding disc locations in the field of view. At this point the command VCALIBRATE V1,V2,R1,R2 is entered on the terminal.

This command sets the vision system coordinate scaling such that the distance between V1 and V2 is equal to the distance measured by the robot between R1 and R2. The vision system calibration need only be done once at system set-up. It can be stored on floppy disc. Note that if the camera is moved or the camera threshold is changed, the system must be re-calibrated.

Camera-to-robot – calibration

Once the vision system has been calibrated, the transformation relating the field of view to the robot must be calculated. This is accomplished by a straightforward procedure similar to that used to calibrate the vision system. Using the pointer and disc, four robot locations and their corresponding vision locations are determined.

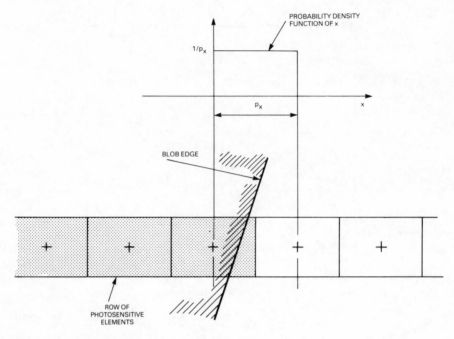

Fig. 4. Uncertainty of edge due to quantisation.

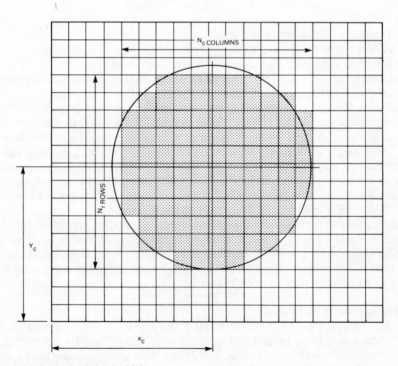

Fig. 5. Centroid of quantised circular blob.

The disc is placed at one corner of the field of view and R1 and V1 are defined as above; then R2 and V2 are defined at an adjacent corner; R3 and V3 are defined at a corner diagonally opposite R2 and V2; and finally R4 and V4 are defined at the centre of the field of view. The user then enters the commands:

DO FRAME TOB = R1,R2,R3,R4
DO FRAME TCB = V1,V2,V3,V4
DO INVERSE TBC = TCB
DO SET TO.CAMERA = TOB:TBC

TO.CAMERA is the transformation which 'takes the robot' to the origin of the camera coordinate system. If the user entered the command DO MOVE TO.CAMERA on the terminal the robot would move so that the pointer was at the origin of the camera coordinate system.

To understand this procedure one must first understand the VAL FRAME and INVERSE instructions. The FRAME instruction defines a transformation which describes the relationship between the robot reference frame and a secondary reference frame or coordinate system. In the example above, the secondary coordinate system is defined to have its origin at R4, its X axis parallel to a line from R1 to R2 and the plane of the Y axis defined by R3. Thus, referring to Fig. 3, TOB may be thought of as an operator which translates and rotates the robot coordinate system to the 'B' coordinate system. Similarly, TCB may be thought of as an opertor from the camera base coordinate system to the 'B' coordinate system (In Fig. 3 the camera base system is denoted by Cx and Cy). Parts will be located by the vision system in camera coordinates. We must get from the robot base to the origin of the camera coordinate system. This is accomplished by inverting the TCB transformation to obtain TBC – an operator from the 'B' coordinate system to the camera base coordinate system. Now we concatenate transformations or use what is called in VAL a 'compound transformation' and define TO.CAMERA as TOB:TBC. TO.CAMERA is then the 'combined operator' which takes the robot coordinate system to the camera coordinate system.

In general, parts will not be located at the origin of the camera coordinate system. The vision system will determine the location and orientation of the part with respect to the camera origin (typically the centre of the field of view). Part offset and rotation are treated as a second transformation which may be concatenated with the first. If we label the robot-origin-to-camera-origin transformation as TO.CAMERA and the camera-origin-to-part transformation as TO.PART (i.e., the location of the part as determined by the vision system) then the instruction MOVE TO.CAMERA:TO.PART would move the robot to the centre of area of the part.

It is often desirable to teach a pick-up point relative to the centre of area of the part. This may be done by defining another transformation which we will call PICKUP. To define PICKUP, the instruction DO APPRO TO.CAMERA:TO.PART, < distance > is issued, where < distance > is the height above the work surface (in mm) necessary to avoid crashing into the part. The robot will move to a position above the part. The robot is then moved using manual control to the desired pick-up location. PICKUP is defined by typing HERE TO.CAMERA:TO.PART:PICKUP. Subsequently, when a part is located by the vision system, the VAL instruction MOVE TO.CAMERA:TO.PART:PICKUP would move the robot to the pick-up point on the part.

The evaluation of compound transformations is a time consuming process in VAL. If a compound transformation is to be used more than once (without any of its component parts changing), it is advantageous to have the compound transformation evaluated once, and assign a name to the resultant transformation. This can be done with the VAL SET instruction, which is an assignment function for location variables.

A sample program is listed below which illustrates how a vision program is written on the VAL language. The instructions which relate specifically to vision are: PICTURE, LOCATE, and FINDHEAP.

PICTURE instructs the vision system to take a picture and process the image.

LOCATE searches the processed blobs in the picture to find any that correspond to the specified prototype and returns its position and orientation. If none are found a branch address is specified.

FINDHEAP searches the current camera image for the largest unrecognised blob and returns its position and orientation.

The sample program takes a picture and tries to find PART1, PART2, or PART3. When a part is found, a motion subroutine, MOVE.PART, is called. MOVE.PART will actually pick up the part and move it to its destination. It is possible that two parts may be touching; this creates an undefined blob for the vision system. After checking for PART3, the program checks to see if there are any undefined blobs by executing the FINDHEAP instruction. In this case, if two parts are touching, they are separated by commanding the robot to sweep its hand through the centre of the unrecognised blob (subroutine STIR).

```
PROGRAM     FIND. PARTS
        REMARK    *** PROGRAM TO FIND PARTS 'PART1', 'PART2' & 'PART3'
        REMARK              AND PUT THEM IN A CHUTE
        REMARK    * RECORD THE CURRENT CAMERA IMAGE
10      PICTURE
        REMARK    * LOOK FOR OCCURRENCES OF PART1 AND MOVE THEM
20      LOCATE    PART1,0,,40
30      SET       X = TO.CAMERA: PART1:PICK1
        GOSUB     MOVE.PART
        GOTO      20
        REMARK    * GOT ALL THE PART1'S, LOOK FOR PART2
40      LOCATE    PART2,0,,60
50      SET       X = TO.CAMERA:PART2:PICK2
        GOSUB     MOVE.PART
        GOTO      40
        REMARK    * GOT ALL THE PART2'S, LOOK FOR PART3
60      LOCATE    PART3,0,,80
70      SET       X = TO.CAMERA:PART3:PICK3
        GOSUB     MOVE.PART
        GOTO      60
        REMARK    * GET HERE IF NO RECOGNISED BLOB SEEN
80      FINDHEAP
        HEAP,0,,90
        REMARK    * UNRECOGNIZED BLOB SEEN, TRY TO BREAK IT UP
        SET       X = TO.CAMERA:HEAP
        GOSUB     STIR
        GOTO      10
        REMARK    * GET HERE WHEN NO MORE OBJECTS ARE VISIBLE
90      TYPE
        TYPE      * NO MORE OBJECTS VISIBLE *
        STOP
```

Accuracy of the Univision system

To evaluate the accuracy of the hand-eye system and to isolate problem areas, the 'locate' procedure was decomposed into steps and tested. The following four steps or components of the locate procedure affect the accuracy of the end result, the accuracy with which the robot hand grasps visually detected objects.

○ VS-100 supplied object locations,
○ arm accuracy,
○ the camera-to-robot transformation,
○ hand position with respect to the object.

What influences the accuracy of the MIC location information? The following items will be discussed individually.

● image quantisation,
● camera geometry,
● threshold,
● calibration.

Image Quantisation. The VS-100 video cameras use a 2-D array of sensor elements, ranging from 128×128 to 240×240 pixels depending on the particular model used. Discrete binary images like these have a worst-case location error proportional to their resolution. The maximum error is a 1/2 pixel shift in a row and a 1/2 pixel shift in a column (Fig. 4). However, the actual uncertainty of the centroid of a blob is generally a small fraction of a pixel. For the centroid of a quantised circular blob, Hill[1] showed that the centroid uncertainty is inversely proportional to the square root of the number of columns or rows in the blob (Fig. 5).

Camera Geometry. The sensor array must be perfectly rectangular (or square in the case of the MIC-22 camera). Further, all sensors must be equally sensitive over an

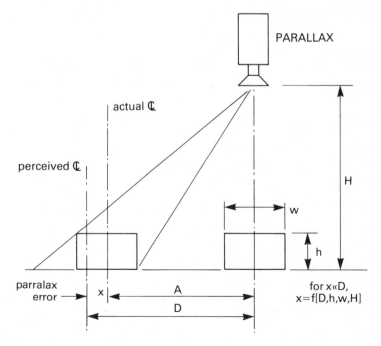

Fig. 6. Parallax error.

equal area. The camera's line of sight must be perpendicular to the table. The lens must be distortion free and accurately focused.

Threshold. The ideal brightness threshold is midway between the brightness of the background and the brightness of the objects to be identified. The choice of threshold within the gap between ranges will influence the accuracy of location computations because of the boundary decisions. When a blob edge intersects a pixel, that pixel will have a recorded intensity between those of the object and the background. The ideal threshold would accurately determine which covers more of the pixel (Fig. 4).

Calibration. The calibration of the VS-100 vision system greatly influences the accuracy of location information. Tests indicate that after following the calibration procedure described earlier, the standard deviation error is approximately 0.5%.

Parallax. The MIC vision system processes only 2-D silhouettes. Depending on the object's height and distance from the centre of view, its silhouette may be formed by its top only or by its top and sides. Visible sides influence the calculation of the blob's centroid, and, therefore, the object's location. Currently, a correction factor for parallax is not included.

Referring to Fig. 6:

Problem: Given H, w, h, and D, find A.

Solution: $A = (4*D*(H-h) - w*h) / (4*H - 2*h)$

Problem: Given H, w, h, and a maximum possible A, find the maximum parallax error: X.

Solution: $X = (2*A*h + w*h) / (4*(H-h))$

Robot Accuracy

Not only must the vision system accurately locate the workpiece, but the arm must go where it is directed. Early PUMA manipulators had a repeatability far superior to their accuracy; accuracy over a typical field of view (300mm × 400mm) might be ±1.2mm. Recent work has greatly improved this accuracy yielding ±0.3mm.

Camera-to-Robot Transformation

How accurate is the TO.CAMERA transformation? That depends, of course, on the accuracy of the vision system and robot locations, and on the care exercised during the defining procedure. Tests indicate between a 0.5% and a 1% error with most of the error due to robot inaccuracy. This should improve markedly with the more accurate robots mentioned above.

Hand Position with Respect to the Object

As there will inevitably be some error due to the above inaccuracies, the hand will not be positioned perfectly with respect to the grasp point. Consequently, grasping the object will move it some small amount. Whether this will introduce additional position error depends on end effector design and part geometry.

Summary of System Accuracy

Several errors can accumulate to introduce inaccuracy to the system. At this time, the largest errors are robot inaccuracy, and parallax shift in the apparent image. The system accuracy can be expected to be approximately 1% of the field of view when the parallax error is small. Present and future work in improving robot accuracy should reduce this number.

Summary

The Univision system is a third generation robot system with integrated visual sensing. The addition of visual sensing allows the robot to acquire randomly located objects. The system is most effective when parts can be identified by their silhouettes and when extreme accuracy in acquisition is not required. The system is easily set-up, trained, and programmed using the high level VAL computer language. The enormous flexibility of a multi-degree of freedom manipulator coupled with a sophisticated vision system offers unprecedented capabilities in the materials handling, inspection, and assembly arenas.

References

[1] J. Hill, C. Rosen, et al., 'Machine Intelligence Research Applied to Industrial Automation', Eighth Report, NSF Grant APR75-13074, SRI Project 4391, SRI International, Menlo Park, California (August 1978).
[2] 'VS-100 Reference Manual', Machine Intelligence Corporation (1980).
[3] P. F. Rogers, 'The PUMA', the VAL Language and Programmable Assembly' (1979).
[4] S. Roth, 'Accuracy of the MIC Vision System', Unimation Document (1980).
[5] 'User's Guide to VAL' – Vision Supplement Unimation Inc. (1980).

S.A.M. OPTO-ELECTRONIC PICTURE SENSOR IN A FLEXIBLE MANUFACTURING ASSEMBLY SYSTEM

W. Brune and K. Bitter, Robert Bosch GmbH, Germany

First published in Feinwerktechnik & Messtechnik, Vol. 90 (1982), No. 2, p. 53 and reprinted by kind permission of Carl Hanser GmbH & Co., Munich, Germany

Abstract

A typical application, namely the flexible, automatic assembly of equipment bearing plates, has been taken to show the practical possibilities of a picture sensor system. It consists of three video cameras, two monitors for picture and data processing with dialogue, a keyboard and a processing unit. The sensor system controls transport, handling and grasping functions and recognises incomplete assemblies in the final check.

Principles and terminology

In medium and small batch production it is important to have flexible, and therefore economical, manufacturing and testing methods. For this reason, we rely upon human beings, with their visual faculties, in such cases and also in large-scale production, where problems cannot be solved with mechanical means. However, there is the desire to close this automation gap with optical sensors, leading to the use of opto-electronic systems for recognition, measurement, testing and positioning tasks. They are created by the connection of a video camera to a high-speed microprocessor system.

An essential condition for the use of optical sensors is that the scene be arranged in a way which is suitable for the sensor. By using suitable lighting and discrimination, a typical binary picture (Fig. 1, below) can be obtained from a photographic grey picture (Fig. 1, above) of the workpieces. The information from the television picture elements, such as area content, contour line length or number of holes in the workpiece, which is reduced to black and white levels in the binary picture, can be used as recognition characteristics of the parts.

The acquired data are filed in a scene table (Fig. 2). In order to recognise workpieces which have already been processed once, it is sufficient to compare these stored data with those being created at the time, taking into account various parameters and processing algorithms. The centre of gravity of the surface of the objects is taken to measure the position of the parts. The rotational position is established by means of the polar check or from the respective position of significant holes in the part. In order to check dimensions of the workpiece, processing algorithms are used which result in a high accuracy of measurement when displaying parts of the workpiece shape.

Fig. 1. Individual parts of a bearing assembly (above: grey picture, below: binary picture. Bearing plate
projection approx. 75 × 40mm).

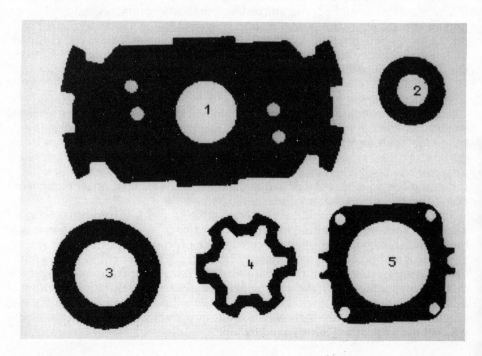

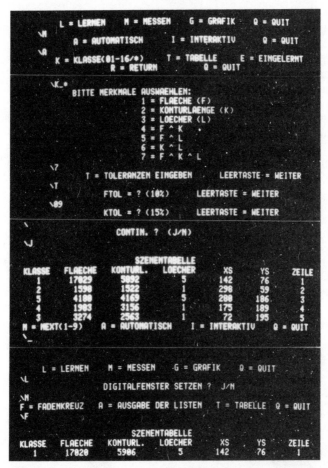

Fig. 2. Documentation of the learning process (upper blocks). Recognition during measurement and scene table with part characteristics (lower block).

Sensor technology and system construction in general

There are two ways of meeting the requirements placed upon an opto-electronic sensor system with picture processing in a two-dimensional cartesian coordinate system. In the one case, where the task is rigidly defined, the 'hardware solution' can achieve the desired results. However, in this case it is virtually impossible to make any changes in the functions, take new requirements into consideration or re-use the system in a similar or different application. On the other hand, easily adaptable, flexible 'software systems' usually operate with a microcomputer and are therefore unable to make use of the advantages of the short processing times offered by hardware equipment. As their processing rate is significantly lower, this often prevents them from being used.

The Sensor System for Automation and Measuring Technology (S.A.M.) shown in Fig. 3, which is of modular construction using hardware and software, achieves short processing times and high flexibility in equal degrees. The microcomputer contained in the basic version can access the picture store and obtain for it the desired characteristics. Additional hardware processors, some of which are used to create the binary picture, can also take over this task (Fig. 3, right). The teach-in set, consisting of data and scene monitors, keyboard and interface modules for communication with a guiding or handling device, complete the system, which is constructed on double European type PCB's in accordance with DIN 41494. The software with the operating system administers the hardware processors, processes the data obtained during assembly of the picture and files them in the scene table (Fig. 2). Time critical, optimised programs are implemented in ASSEMBLER, other structures in the higher programming language PLZ.

In order to recognise workpieces it is sufficient to compare the data, which have already been learned in the teach-in process by means of operator guidance or synthetically predetermined, with those which are established automatically during the measuring process (Fig. 2). Various algorithms are available in the software library for picture processing and measuring tasks. When using both picture stores, with 256 × 352 pixels each, and picture refinement (specific picture transformation) even scenes which have been impaired by small foreign bodies (shavings, splinters, etc.) can be processed properly.

Suitable illumination of the object scene must ensure that errors cannot occur as a result of accidental shadow formations, reflections or disturbing outside light. The scene illumination should be such that it dominates the surrounding light. There is a wide range of different types of illumination: trans-illumination, direct illumination, flash light (from a stroboscope or infra-red light), fluorescent

Fig. 3. S.A.M. opto-electronic sensor system. The basic development was done by IITB Karlsruhe, Robert Bosch GmbH, Television Systems Division, Darmstadt, have done further development and are now manufacturing it for industrial applications.

Fig. 4. System for joining bearing parts (left the opto-electronic sensor, above left camera 1, right
camera 3).

radiation, polarised light, coherent light or light section method, etc. The distance
between the camera and the object and the angle of view should also be taken into
consideration, in order to avoid perspective distortions.

The picture section must be selected to meet the required measuring accuracy,
whereby the geometry error shows by what percentage of the picture height B_H the
video picture points are displaced form their ideal position in the worst case. For
this reason, the image of objects which are to be processed by sensors should fill the
format as far as possible so that the measurement error does not become considerably greater than that of the television camera. (Measurement camera $\pm 0.5\%$ B_H,
standard cameras $\pm 1\%$ B_H within a measurement circle of 90% picture format of
4:3 (width: height) laid down in CCIR. These data are decisive in the choice of pick-
up tube for the video camera, whereby the state of motion of the object still has to be
clarified. With stationary objects no distortions are caused by the object moving
within the TV camera's scanning time of 20ms per frame (50 frames per second).
With slow movements and low accuracy requirements the resulting lack of
definition caused by movement is negligible. Parts which move at a faster rate
would result in errors. They are therefore illuminated by flash light in order to
'freeze' their movement optically for the camera's eye.

Use in flexible assembly

Fig. 4 shows a flexible, automatic assembly system in which the five individual parts (Fig. 1) of an equipment bearing are put together[1]. The individual parts and the completed assembly are transported in order in magazines or on pallets and joined on specially adapted workpiece supports. The joining and transporting processes are monitored and checked by the sensor system described in this article, with a final check as to whether assembly is complete. The joining and handling processes are numerically controlled (PU 4000 in Fig. 5) in four axes (two slide rails, one linear unit and one rotation unit); a PC 4000 control, with a programmable memory, also supplied by Robert Bosch GmbH, Industrial Equipment Division, monitors the whole systems, controls the circulation of the workpiece supports in the working area and the feed device for the individual parts and informs the sensor system which measurement point is to be called up. The connection between the NC axes and the sensor system consists of one serial data line respectively.

A total of 12 different scenes are evaluated. In addition to the operations required for picture processing, such as altering the binary coding threshold, the use of picture refinement or the recognition algorithms, it is necessary to prepare the acquired data for the handling system. This involves a special type of data transfer and scale conversion from pixels of the binary picture into the measurement system of the handling device.

In view of the light surface of the parts and the matt black pallets, 900 Lux illumination density is sufficient to create an adequate contrast. Four 60W fluorescent tubes were used in addition to the factory lighting.

The bearing plates, which are the largest parts (approx. 40 × 75mm), arrive on pallets, in no particular order. The gripping device in the assembly system can only take up these parts from a certain position range, which makes it necessary for them to be pre-sorted. Otherwise there is no restriction as regards the rotational position of the parts on the pallet; as a result of the shape of the gripping device and the recognition and positioning characteristics used, they are even allowed to touch

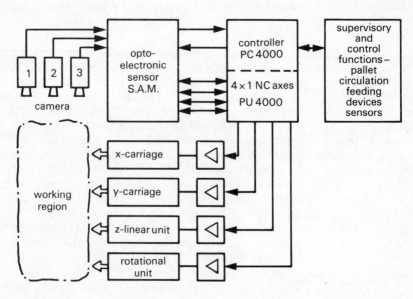

Fig. 5. Control of the assembly system.

Fig. 6. Measurement of the bearing plates lying unordered on the pallet (above: coarse measurement, below: fine measurement).

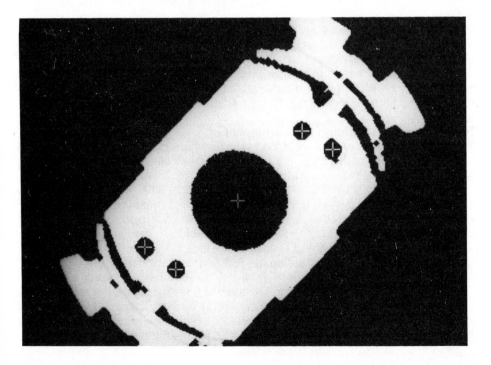

each other. In order to be able to create a sufficient supply of parts, several pallets are stacked on top of each other, which are handled by a second gripping device. The sensor system also monitors the automatic change-over between gripping devices and pallets. In order to carry out any scale correction which may then be necessary, the sensor registers marking points on the pallets. The stacks of empty pallets stand directly next to the filled pallets.

Both areas cover 400 × 600mm and are observed by a camera at a distance of 2.4m, using a lens with a focal length of 36mm. This enables positioning measurements of the parts to be photographed to be carried out with a measurement error of ± 4mm. As the joining accuracy has to be ± 0.2mm it is essential to carry out precise measurement with an additional, portable camera. Given the position from the coarse measurement, the precision measuring camera is brought near to the part on the arm of the robot. The precise measurement then provides exact data for joint accurate gripping and setting down on an assembly block.

In the further course of the circulation of the workpiece the spring washers and cage parts are released from vibration conveyors and taken out from defined take-up positions. The sensor system is not involved in this stage.

A third, rigidly mounted camera (distance 2400mm, image field 300 × 400mm, lens focal length 50mm) observes the supply sources for the sliding bearings and lubricating felts, whereby not only the presence of the felt rings is checked but also their position is measured.

The joining together of all the parts is followed by the final check. Once the first camera has established that there is a free space on the stack of empty pallets, fully assembled parts are deposited on it. The handling system throws faulty parts on to a third pallet. For checking the completeness of the parts, in order to obtain characteristics from the binary picture which describe the status of the assembly, algorithms were developed with the aid of picture processing operations and automatic binary coding threshold. The processing point in question is selected via picture windows and picture information which is not needed is thus screened out.

Operational process

In the course of the joining process twelve different scenes are analysed in order to process numerous individual tasks. They range from registering the scale, as the height of the pallets varies, to establishing the position of the individual parts and checking the completeness of the assembled parts. The tasks at three operating positions shall be explained as examples:

Recognition of the position of the bearing plates

Fig. 6, above, shows the scene recorded by the first camera, where the coordinates of one part are marked with a cross. This workpiece is selected by the sensor as having the shortest handling path out of all the parts measured.

After the position coordinates of the sensor system have been converted into those of the handling device, the precision camera mounted on the gripping arm travels across the part to the established approximate position. As this camera is approx. 150mm above the object, the image field is correspondingly smaller (approx. 60 × 46mm) or the object is shown to be larger. The resulting measurement accuracy (deviation from the coordinates ±0.2mm, rotational position error ±1°) allows the part to be handled while maintaining the required joining accuracy.

The details shown in Fig. 6, below, are taken to form the result. For this purpose, the control process filed in the software inverts the binary picture . The holes in the workpiece now appear as individual objects, of which the surface centre of gravity coordinates can be established. After a total of 150ms the position and rotational

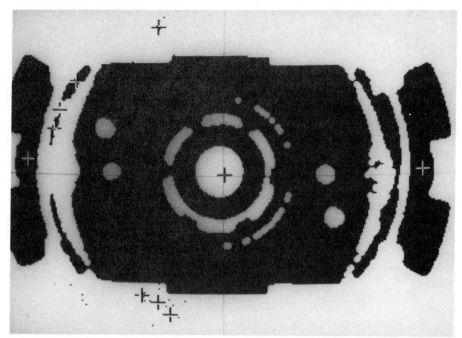

a. Binary picture of complete assembly before picture processing operations.

Fig. 7. Final check as to completeness.

b. Binary pictures after picture processing.

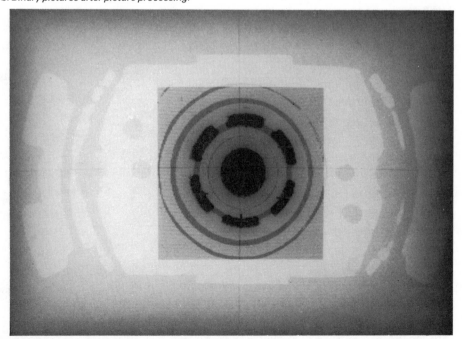

b₁. Complete assembly.

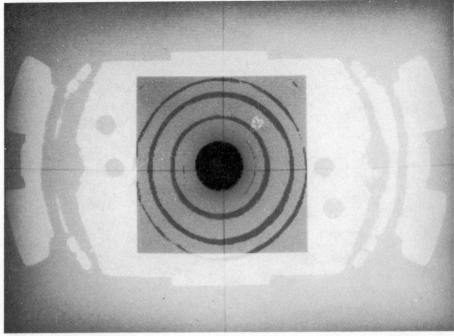

b_2. Spring washer missing.

Fig. 7b contd.

b_3. Felt ring missing.

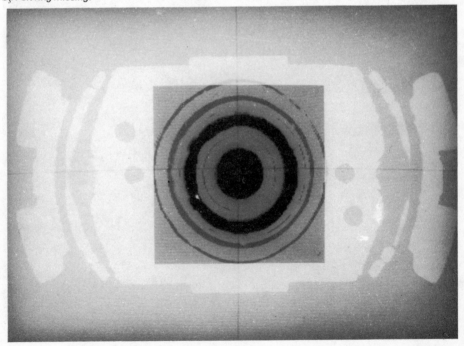

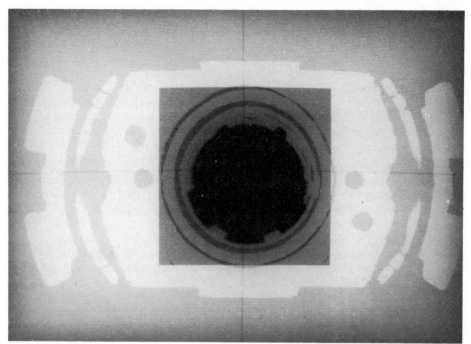

b_4. Sliding bearing and felt missing.

Fig. 7b contd.

b_5. Cage missing.

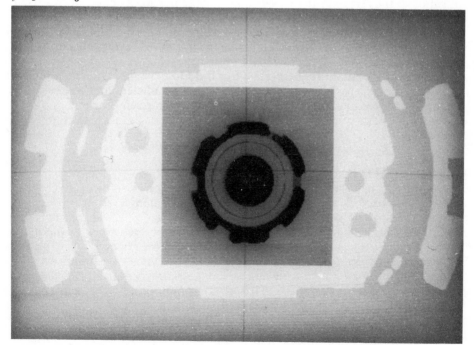

position data are transferred, converted to the handling device's measuring scale. The part can now be picked up by the gripping device and placed in the assembly position.

If the sensor system recognises that some parts are missing on the pallet, the assembly system is given the command to change the pallet. First, the other gripping device automatically comes into operation, which places the emptied pallet on the stack for the completely assembled parts. Then the first gripping device is reactivated and the next parts are taken from the pallet underneath it, in the manner described above. An indication is given once all the pallets have been emptied; the system is activated again with full pallets.

Monitoring the sliding bearing dispenser

The master control PC 4000 has to be informed whether parts are still available in the dispenser magazine for the sliding bearings. In view of the low contrast between the background and the workpiece, the binary coding threshold has to be automatically adjusted according to the area criterion, in order to avoid measurement errors as a result of fluctuations in brightness. The sensor system can always recognise with certainty whether there are still parts left in the magazine or whether the refill function should be activated. If no more sliding bearings are fed into the magazine, a display appears on the data monitor indicating 'no sliding bearings present'. The system then comes to a temporary halt. Once the dispenser magazine has been filled up the assembly system proceeds to the next stage.

Final check as to completeness

Fig. 7 shows how checking criteria for the object to be processed can be gained from the scene by picture processing. The number of area points in Fig. 7a shows whether a part with a base plate is present at the checking point.

It is more important, however, that the other parts (lubricating felt, sliding bearing, spring washer and cage part) should be assembled. Thus, in addition to activation of the final riveting process, there is the possibility of ensuring selected materials are sufficiently stocked.

The picture processing operations erosion and dilatation enable Fig. 7b to be produced, which shows a fully assembled part. It is possible to see from the inner surface whether the sliding bearing is assembled; the six areas, which are situated on a concentric circle, indicate the presence of the spring washer and felt, which are covered by the cage part. Trials have shown that it is sufficient for three surrounding areas to be recognised for a reliable indication to be given. The binary pictures of incomplete parts (Figs. 7b to b$_5$) give a clear indication as to which components are missing.

Operational reliability

The sensor system has test routines, which are backed up by software, to check the hardware functions. In the case of mains failure, the data in the scene progam table are stored in a battery-buffered store for up to 120 hours. When the system is starting up, the sensor system tests its own interrelated processes before the master control is given the signal to enable the start. The data are checked for plausibility while being transferred. The adjustment and measurement accuracy are tested through establishment of the scale and simultaneous checking of the height of the pallet stack by inductive sensors. The binary coding threshold is adjusted with respect to time and the type of scene. This has made it possible to achieve practically one hundred per cent operational reliability. To date, it has not been possible to locate any errors which could be clearly attributed to the sensor.

Summary

Opto-electronic sensors with binary picture processing and access to reduced data derived from a scene are used to analyse one or several workpieces in one image field. The characteristics of the system create the prerequisites for measurement and recognition tasks as part of position registration and determination of the rotational position, as well as for quick and simple change-over to new workpieces in the teach-in process. Surface checks require high resolution systems, which will be the subject of future development. Selection of the video camera's pick-up tubes depends on the type of illumination and the movement of the object. The part of the scene displayed and the accuracy of the sensor result determine the camera and lens type. The environmental conditions and algorithms relevant to the task to be performed determine the extent of the hardware and the software.

For 'learning' purposes the parts are presented in the image field of the camera. Workpieces which are designed with the sensor in mind and illuminated to provide a good contrast thus allow further steps towards automation in industrial manufacture.

Reference

[1] *Feinwerktechnik & Messtechnik*, 89 (1981), p. 413.

VISION SYSTEM SEPARATES GATHERING FROM PROCESSING

G. I. Robertson, Control Automation Inc., Princeton, USA

Reprinted from Sensor Review, Vol. 2, No.2 (April 1982)

Abstract

The V-1000 is an intelligent binary system from Control Automation Inc., which employs a novel hierarchical control to separate the functions of data gathering and data processing. It uses capabilities of commercially available computers, and offers a high performance at a low cost.

Introduction

A high precision cartesian assembly robot and an intelligent vision peripheral are the first two products of Control Automation Inc. They were introduced at the Robots VI Show in Detroit, March 1982. Unique architecture and the optimal use of distributed processing and state-of-the-art electronics, has led to the price of both these machines being competitive – the Vision System is $19,700 complete, and the Assembly Robot starts at $69,000 for a one arm machine and $13,000 for each additional arm. A two armed robot complete with an integrated vision system less than $100,000.

Both machines were constructed using the concept of distributed processing and hierarchical control. In the case of the vision system, there are two levels of control. The vision processor itself – the V-1000 – performs data gathering, image analysis, part identification and orientation.

The V-1000 is driven by an external computer, usually the system controller SC-1000, but it may be any computer at all. The function of the external computer is to command the V-1000 to gather data then to record and make decisions based upon this data.

The vision system is ideally suited for either stand-alone applications or as a robot peripheral.

Function

The V-1000 vision processor performs parts recognition and location, inspection and measurement on non-overlapping objects which are illuminated in such a way that they provide a high contrast between themselves and their background. To accomplish this the V-1000 uses an algorithm first developed by the Stanford Research Institute, known as the 'SRI algorithm'.

The fundamental process of the SRI algorithm is to segment the picture presented to the camera into non-overlapping separated objects. For each object, several parameters are then calculated which are descriptive of the object but not affected by its position or orientation. Such parameters are area, perimeter, second moment about the centroid, etc.

In the V-1000 approximately nine such parameters are measured and these are used as descriptors for the object. During teaching the object is presented to the camera in several different positions and orientations, and the V-1000 is instructed to gather data and mark the data with the name of the object. Later, when asked to recognise an object, the processor calculates the characteristic data and then matches this information against the taught data.

Fig. 1 is a block diagram schematic of the V-1000. Up to four cameras may be used; each is a solid state 128×128 element camera which is driven by an analogue interface board. The data are digitised and thresholding is achieved by comparing it against a preset level to convert the picture into a white on black binary picture. The thresholding level may be set by a software instruction and may optionally include some hysteresis.

After thresholding, the data are compressed to select only information of significance. This data are then analysed by the central processing unit, an Intel 8086 based system operating at 8MHz for maximum speed. Critical parts of the code are written in assembly language to further speed the operation.

Control
The V-1000 is configured strictly as a peripheral, and to the controlling computer it appears electrically identical to a computer terminal. Communication between the V-1000 vision processor and the user is through an RS-232C (or optional RS-422) standard serial communication line. This communication protocol was selected because it is the only interface that nearly all computers are equipped with as standard, and no extra interfacing is needed. The time taken to communicate

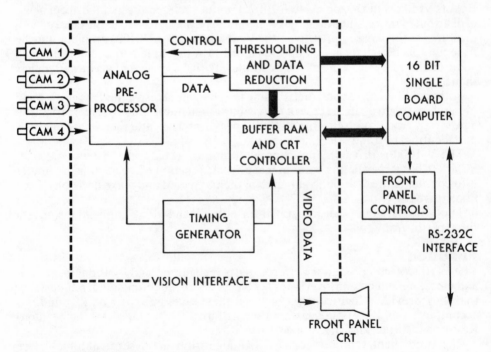

Fig. 1. V-1000 Vision processor – block diagram.

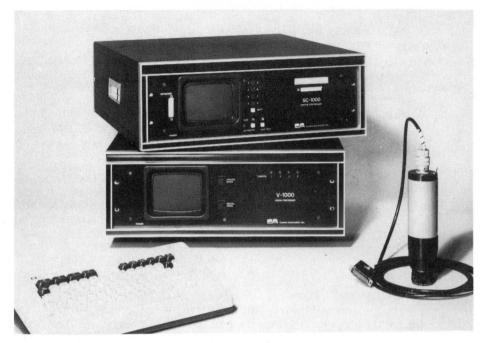

Fig. 2. V-1000 Vision processor with SG-1000 system controller.

between the V-1000 and the external computer is only a few milli-seconds when running at 9,600 Baud, adding no significant overhead to the overall cycle time.

The V-1000 is also configured as a peripheral in the operational sense, and makes no decisions or value judgements itself. It operates strictly as a data gatherer in response to commands from the external computer that does all process control.

This separation of functions markedly increases the flexibility of the system since the controlling computer may be selected according to the job to be performed. A low cost simple computer could be chosen for straightforward jobs, or a large main frame computer could be selected to run many vision systems simultaneously.

A low cost but capable computer ideally suited to running the V-1000 is the HP 85 desk top computer, which is incorporated (in its 9915 industrial version) in the SC-1000 system controller. The SC-1000 is supplied with the vision system when purchased as a complete unit.

Commands

The V-1000 responds to approximately 20 different commands, a few of which are listed in the table. These commands are passed down the serial communication line from the SC-1000 system controller or other controlling computer in the form of standard ASCII strings exactly as if writing to a terminal or printer. For example, the command – PICTURE, 2 – instructs the V-1000 to take a picture through camera 2 and to analyse and store the results. A later command – WHEREIS, WIDGET – instructs the V-1000 to analyse this stored data and to return the X, Y, theta and parity coordinates of the object which had been previously taught under the name 'Widget'.

A powerful feature of the system is the MEASURE command. When this command is issued, followed by the X and Y locations of two points on the screen,

SAMPLE COMMANDS AND RESPONSES

Command	Response	Intepretation
Picture 1		Take a picture on camera 1 and interpret the image
Shownext		Display only one of the objects from the current picture
Learn, Widget		Learn the features of the currently displayed object and describe them as 'widget'
Whereis, Widget		Locate an object matching description of 'widget' and return X, Y (–) and Parity
	24,82,114,0	Centroid location is 24 pixels, 82 pixels. Orientation 114°, Not inverted.
Measure, 22,48,104,90		Draw a line between (22,48) and (104,90) and return white length, black length, and number of transitions.
	31,61,3	Length of white (object) was 31, length of black (background) was 61. There were 3 transitions

the V-1000 returns three numbers – the white length of the line joining the two selected points, the black length along the line and the number of transition points (white/black or black/white) encountered along the line. Since the position of the object has been previously found, the coordinates of the points to be measured may be readily calculated. This unique feature allows detailed measurements to be made on any part of a randomly oriented object.

The full power of the controlling computer allows detailed analysis of the data to

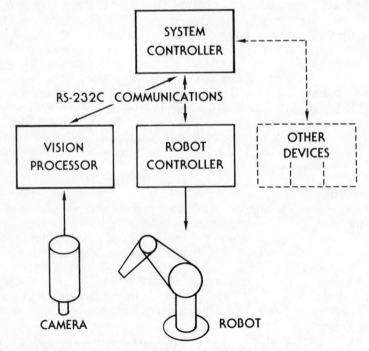

Fig. 3. *Vision and robot systems controlled by the same controller.*

be made. For example, a simple function such as counting the number of 'widgets' could be performed, or a complex function such as measurements of certain holes and statistical computations of their discrepancies could be made.

The SC-1000 system controller is programmed in Hewlett Packard's industrial BASIC, which is further enhanced by key-words that reduce the programming work for user. BASIC was selected to avoid the proliferation of special purpose languages and in the belief that all engineers who are familiar with any programming are comfortable in BASIC. Since the SC-1000 is only responsible for putting together and issuing a command to the V-1000, the relatively slow BASIC language does not in any sense limit the capability of the system.

Of course, since any computer may be used in place of the SC-1000 it is obvious that any language may be used to put together the command and interpret the answers. Thus a company with extensive Digital Equipment Corporation equipment and much experience in FORTRAN may prefer to use a PDP 11 operating in FORTRAN. The V-1000 will work just as well.

Fig. 2 shows the vision system package including the V-1000, the SC-1000 system controller and one camera. This capable and low cost system is complete and ready to be switched on.

Performance

An early design goal of the V-1000 was high speed. An on-line vision system is not useful if it takes significantly longer than one second to process a picture, so a design goal of 0.5s was fixed for simple pictures. Of course, since the analysis time is a function of the complexity of the scene the time increases for more complex pictures, but typical analysis times of 0.5 to 0.75s are obtained.

The accuracy of the system is 1% of the field of view. This means that an object 10in. in diameter may be measured to 0.1in. However, for gauging or other accurate work, two cameras could be employed, each one looking at close range at opposite edges of the object. Since the cameras are separated by a known distance, the controlling computer may easily offset the readings of the two cameras and so obtain a high resolution measurement.

Although the separation of functions between the vision processor and system controller generates great advantages for stand-alone vision systems, an even greater benefit is obtained because of the ease of interfacing to other systems, especially to a robot.

Fig. 3 shows a vision processor being controlled by the same controller that is driving a robot. Because of the RS-232 interface protocol, the interfacing is a straightforward plug-in operation. When the robot controller wishes to know the position of an object for acquisition, it merely instructs the vision processor to take a picture, then issues a 'WHEREIS, OBJECT' whereupon the coordinates of the object are returned on the serial communication line. A robot controller with mathematical capability may easily transform these numbers into robot coordinates by simple multiplications. Alternatively if the robot controller does not have this capability, the SC-1000 could be used between the V-1000 and robot and this would convert the data from camera coordinates to robot coordinates.

WESTINGHOUSE GREY SCALE VISION SYSTEM FOR REAL-TIME CONTROL AND INSPECTION

Arkady G. Makhlin and Glenn E. Tinsdale, Westinghouse Electric Corp., USA

Abstract

A 64 grey level vision system has been developed for real-time robot control and inspection. It is capable of processing up to 12 TV frames per second which is an order of magnitude faster than any commercially available grey scale vision system. The system consists of a programmable pipelined hardwired preprocessor for image conditioning, filtering, edge extraction, segmentation, and blob tracking and orientation, as well as for interpreting visual data output for a robot or external systems control and communications. The set-up for assembly of an electro-mechanical relay under vision control is described and the results are presented.

Introduction

Current robot vision systems rely primarily on the processing of binary images; that is, a threshold is applied to the output of the camera so as to provide a black and white image. In this image a correctly illuminated object will appear as a silhouette on a uniform background. For simple configurations, the binary image offers easy determination of parts measurements and locations, and a minimal processing load. In more complex configurations, however, the binary image is inadequate because:

1. Closely spaced parts are joined together;
2. Colouring of parts and background does not permit separation by a unique threshold;
3. Available illumination does not permit a clean separation of part and background.

For these reasons it is desirable to make use of the density information available from most cameras. With this information, as well as the ability to extract such image features as edges and 'blobs', operations can be carried out on complex scenes under difficult and variable lighting conditions.

The Westinghouse Grey Scale Vision System offer grey scale processing at real time video rates.

Hardware and operation

The output of the TV camera, as provided for visual inspection and control, usually consists of a uniform two-dimensional array of density values (pixels). Useful information, such as the outline geometry of objects of interest, must be extracted by computations involving both pixel density and geometry. When these computations have been performed, there is the prospect that the desired information can be expressed at a greatly reduced bandwidth (one or more orders of magnitude) from the original image data.

As a result of the preprocessing operation, we would like to have a collection of image primitives; that is, basic image decriptors, along with their specifications and location in the image. The primitives should be easy to extract, should be consistent in value in spite of noise and should be simply stated, so as to maintain a low data bandwidth. Furthermore, they should be useful for a variety of applications. A primitive is an information package extracted from the raw image data. It should serve as a simple building block for later image operations.

For the operations described in this paper, only two primitive geometric features are used. The first feature is a straight line contour of density gradient (an edge). Edges are defined by their end-points and polarity (relationship of light and dark sides). The second feature represents a small region in the image which is lighter or darker than its surroundings. This is termed a blob. Blobs are defined by their location, their height and width, and their relative density with respect to the background.

Since the extraction of primitives is common to many sensors and applications, and since it results in the compression of high bandwidth image data, it is a logical candidate for incorporation into special-purpose digital hardware. For many applications, the data bandwidth reduction of the digital image preprocessor permits further operations to be performed in a programmable general-purpose processor, with the advantage of great flexibility in the range of problems handled. A Westinghouse grey scale vision system (Fig. 1) includes the image preprocessor and the programmable image postprocessor. The image preprocessor consists of a pipeline series of computations operating throughout the image on 4 or 5 neighbouring pixels. So that computations can be performed in both dimensions, lines of data across the image are stored in a series of shift registers. The computations include the extraction of density gradient, gradient thinning, and the tracking of similar gradient characteristics from point to point to extract edges and blobs.

Details of the grey scale vision are shown by Fig. 2. Interfacing devices, such as the TV camera and display, as well as a robot, are shown at the left. Following synchronisation and digitising of the incoming video signal the data is entered into a full-frame buffer and then extracted from the buffer on a line-by-line basis for entry into the image preprocessor. The image preprocessor accepts a standard RS-170 TV signal.

The configuration and adjustment of the image preprocessor is under software control on a frame-to-frame basis. A variety of pixel conditioning options are offered, including averaging filters, a median filter, level slicing, and rate compression. Level slicing at two levels is provided so that simultaneous isolation of both light and dark regions may be accomplished for dynamic threshold adjustment. Rate compression reduces the flow of image data by selection of alternate pixels.

Pixel conditioning is followed by gradient extraction to obtain both amplitude and direction for a two-by-two pixel array. When this operation is preceded by averaging filters, the result is equivalent to larger operators with more smoothing but lower resolution. The gradient operation is a first derivative computation on the image idea. All pixels with gradient values above an adaptively selected amplitude threshold are processed by the maximise function.

The blob tracker detects conditions which indicate the top of a potential blob. Left and right outlines are then tracked through the image on a line-by-line basis.

Fig. 1. General view of the grey scale vision system.

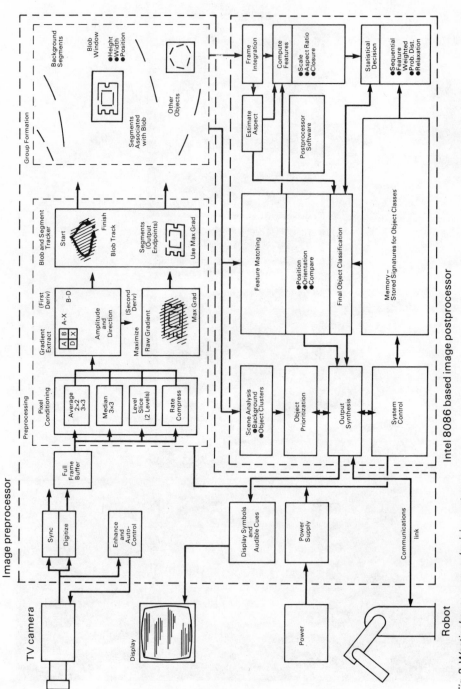

Fig. 2. Westinghouse grey scale vision system.

The length of the vertical path is retained, as well as the extent of the left and right excursions. The blob is detected when a matching left and right outline connect.

Maximise is a second derivative operation which thins the gradient into the fine lines of the target, including both the outlines and the internal detail. It compares neighbouring gradient values in both the x and y directions. In the process, edge sharpness and amplitude are also derived.

The segment tracker connects pixels with compatible features defining the endpoints and gradient direction of image edges. This will occur for exterior edges and interior contrast lines associated with objects, as well as for background definition.

Group formation is dedicated hardware which reduces the post processor load by collecting all data associated with each object. This sorting function is assisted by indentifying tags which are applied to blobs and edges as they are generated. Output from the group formation logic is provided to the image postprocessor. Outputs include groups which represent objects (work pieces) and edges associated with background.

Operations within the image postprocessor, which are fully programmable, may include:
○ Frame-to-frame computation,
○ Feature computation,
○ Statistical decision making,
○ Feature matching for final classification,
○ Scene analysis,
○ Object prioritisation,
○ Output synthesis,
○ System control.

Detailed operations of the preprocessor are shown by Fig. 3. The 50 × 50 array of density values is shown at the top. For ease of interpretation, only 16 grey scale levels are printed. Of the 16 levels, values above 9 are indicated by an overprinted slash (/). Gradient directions are also quantised to 16 directions, with a l indicating a change from light to threshold (the sensitivity threshold on the system), no output is shown, and the location is marked with a dot. The result of a line-thinning operation is indicated at the left, where some of the previous gradient outputs have been deleted if they are adjacent to a similar output of larger amplitude, and if their presence in edge formation is unnecessary. Edges are formed by tracking the thinned gradient outputs, and by recording both the beginning and endpoints of each track. They are shown on the graphical plot at the bottom of Fig. 3. The window at right indicates the accumulated angles associated with tracking around a blob. The resulting blob symbol, a cross, is shown on the graphical plot, along with the edges. The height and width of the blob are indicated by the arms of the cross.

Vision system software

The vision system software has been written generally in PL/M using structured-programming concepts. It has been written as system-independent as possible with the exception of disk and terminal I/O, where operating system service routines are used. The basic interactive routines between the image preprocessor and the post-processor are programmed in the Intel assembly language. The structure of the software is shown in Fig. 4. This software has been written on an Intel 8086 based general purpose computer running under RMX-86, utilising floppy disk as the system's device and primary backup.

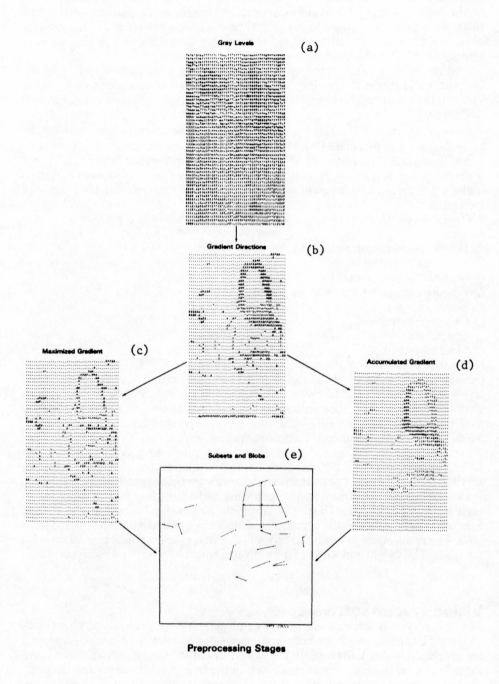

Preprocessing Stages

Fig. 3. Steps in the preprocessing of an image.

Command String Interpreter (VISCSI). This module controls communications with the operator and controls the execution of all other modules. It calls on the image-processing System Nucleus (SYSNUC) which controls the interface to the image preprocessor.

System Nucleus (SYSNUC). Commands the image preprocessor to either:
1. Take a picture and read a preprocessed camera image to interface buffer;
2. Take a picture and read the buffer to the postprocessor memory;
3. Read the buffer to the postprocessor memory;
4. Take a picture, read the buffer to the postprocessor memory, and calculate base feature values;
5. Calculate base feature values from a given buffer;
6. As 4. and add evaluation of the main object.

Feature Selection Module (VISFEA). This module, based on operator input, selects feature value calculation routines from a library of such routines and controls the set-up of working storage for each feature selected.

Recognition Training Module (VISREC). This module is called to establish a set of feature statistics for a part. The same object is presented in the vision field several times and the results of feature analysis are accumulated as statistics. Object recognition can be tested; now, the features of a presented object are verified against the statistics, and the program returns a message whether the object is, or is not, recognised.

Production Training Module (VISTRA). This module is the responsibility of the user-programmer preparing a production program. It serves for training and refers both to system utility routines and to routines specific to the application under construction. Essentially, it differs from the production module only in that it

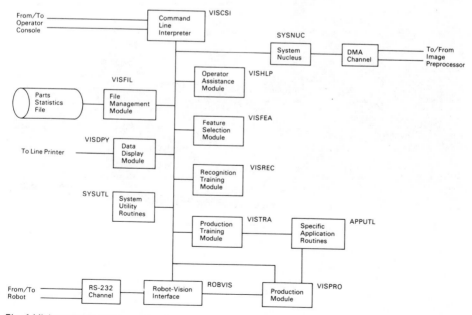

Fig. 4. Vision system software structure.

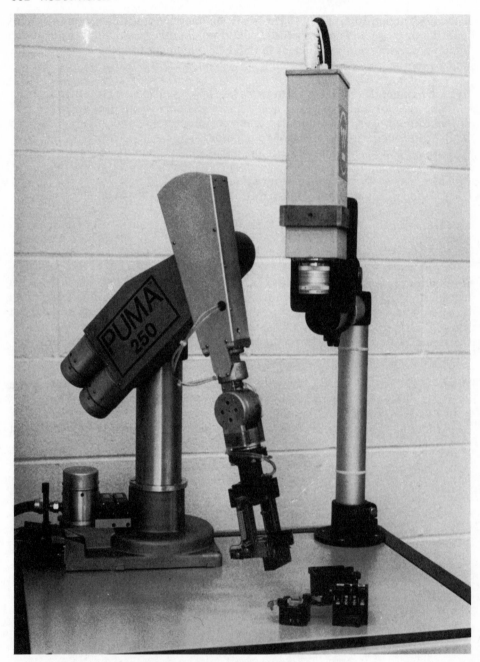

Fig. 5. Assembling an electromechanical relay under vision control.

simulates robot commands through operator input, and outputs to the console the responses that will be sent to the robot. It can also serve to store in memory certain part-specific parameters or their retrieval addresses.

Production Module (VISPRO). This module, like the preceding and the following ones, is the responsibility of the user-programmer. Those three modules constitute what is here named 'specific application programming'. The vision system is thus adaptable to the varied applications of the robot. The Production Module receives its commands from the robot through the robot-vision interface, ROBVIS.

Specific Application Utility Routines (APPUTL). The routines of this module are those subdivisions of the specific programming that are common to the training and production modes.

Robot-Vision Interface (ROBVIS). Entering this module causes the vision system to switch to a passive state, waiting for commands issued by the robot. These commands are decoded and transmitted to the production module for execution and response. The ROBVIS module has access to all the specific application routines of the production test module, and to all those of the system nucleus and system utility modules. This non-interactive robot-vision interface only returns

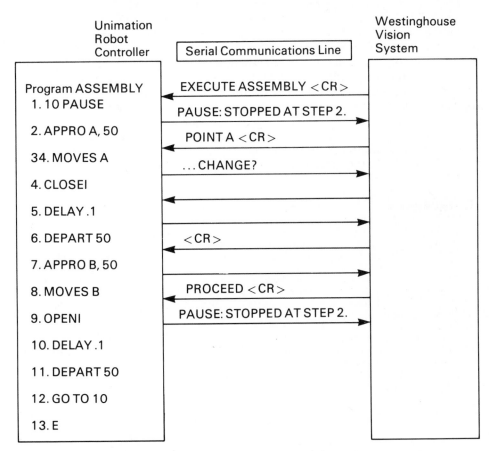

Fig. 6. Vision–Robot communications protocol and a PUMA assembly program.

control to the interactive command-string interpreter when a recognised error has been detected by the vision system but has not been processed successfully by the production module. Such an 'error' can be a specific code transmitted by the robot to release the vision system to its own control. The ROBVIS module is obviously different for different robot-vision systems.

System Utility Routines (SYSUTL). This module has many entries: all the vision system routines accessible both in training and production modes.

Assembly under vision control

The set-up for assembly of an electromechanical relay under vision control is shown in Fig. 5. The PUMA 250 industrial robot was used to assemble a four-piece electromechanical relay. All workpieces were located in the field of view of a Westinghouse TV camera without overlapping with spaces between workpieces. Vision-robot communications were realised through the RS-232 serial link. A PUMA assembly program and a vision-robot communications protocol are illustrated by Fig. 6. The program ASSEMBLY has been written in VAL robot programming language, and the PUMA robot was communicating with the image postprocessor through its controller's CRT port as if an operator typed VAL instructions in the CRT keyboard.

The vision system commands the robot to start ASSEMBLY program (EXECUTE ASSEMBLY). The program stops at step 2, and, at this point, the robot is being asked to display a point A (an arbitrary location of workpiece within the field of view). The vision system, then modifies the location A by sending to the robot values of its position (X, Y, Z) and orientation (O, A, T), calculated on the basis of visual information. Consequently, the robot moves into the newly defined location A, picks up a workpiece and brings it to the part predefined assembly point B. The program returns then to step 2 and the robot receives the next location A, and so on.

The use of the Westinghouse grey scale vision system allows assembly of combinations of similar parts under poor lighting conditions. It is also suitable for real time inspection, where complex parts configurations exist, and lighting is restricted.

References

[1] Pratt, W. K., *Digital Image Processing*, John Wiley (1978).
[2] Rosenfeld, A., Kak, A. C., *Digital Picture Processing*, Academic Press (1976).
[3] Makhlin, A. G., 'Westinghouse Visual Inspection and Industrial Robot Control System', *Proc. 1st Int. Conf. on Robot Vision and Sensory Controls,* Stratford-upon-Avon, UK, IFS (Publications) (April 1981).
[4] Tisdale, G. E., 'A Digital Image Processor for Automatic Target Cueing, Navigation, and Change Detection,' SPIE, vol. 101, Airborne Reconnaissance – Tactical/Real Time (1977).
[5] Helland, A. R., Tisdale, G. E., 'The AUTO-Q Digital Image Processor for Autoprocessing of Reconnaissance Images', Proc. NAECON '80, (1980).
[6] Tisdale, G. E., 'Automatic Registration of Points in Two Separate Images', US Patent No. 3,748,644, July 24, 1973.
[7] Tisdale, G. E., 'Preprocessing Method and Apparatus for Pattern recognition', US Patent No. 3,636,513, January 18, 1972.
[8] Pincoffs, P. H., *et. al.*, 'Classification Method and Apparatus for Pattern Recognition', US patent No. 3,638,188, January 25, 1972.

Authors' organisations and addresses

Mr. T. Bamba,
Mitsubishi Electric Corporation,
Product Development Lab.,
80 Nakano, Minami Shimizu,
Amagasaki, Hyogo 661,
Japan.

Mr. Richard Baumann,
Foundry Plant Engineering,
General Motors of Canada Ltd.,
St. Catherines Plant,
Box 3002,
570 Glendale Avenue,
St. Catherine's, Ontario L2R 7B3,
Canada.

Mr. K. H. Bitter,
Robert Bosch GmbH,
Television Systems Division,
Robert-Bosch-Strasse 7,
6100 Darmstadt,
Germany.

Dr. Robert C. Bolles,
Snr. Computer Scientist,
SRI International,
333 Ravenswood Avenue,
Menlo Park, CA 94025,
U.S.A.

Mr. Brian R. Carlisle,
General Manager,
Unimation Inc.,
West Coast Division,
1202 Charleston Road,
Mountain View,
California, 94043, U.S.A.

Dr. W. Clocksin,
Robot Welding Project,
Department of Engineering Science,
University of Oxford,
Parks Road,
OX1 3PJ.

Dr. A. J. Cronshaw,
Patscentre International,
Cambridge Division,
Melbourn, Royston,
Herts. SG8 6DP.

Mr. M. J. Dunne,
Vice President, Engineering,
Unimation/Kawasaki Joint
Development Team,
c/o Unimation Inc.,
Shelter Rock Lane,
Danbury, Conn. 06810,
U.S.A.

Professor O. D. Faugeras,
INRIA, Domaine de
Voluceau-Rocquencourt,
B.P. 105,
78153 Le Chesnay Cedex,
France.

Dr. M. Gini,
Politecnico di Milano,
Istituto di Elettrotechnica ed Ellettronic,
Piazza Leonardo da Vinci 31,
20133 Milano,
Italy.

Dr. G. J. Gleason,
Manager, Vision Engineering,
Machine Intelligence Corporation,
330 Potrero Avenue,
Sunnyvale, CA 94086,
U.S.A.

Professor W. B. Heginbotham,
Director General,
PERA,
Melton Mowbray,
Leicestershire.

Mr. J.-P. Hermann,
Direction des Techniques Avancees en
Automatismes,
Regie Nationale des Usines Renault,
92109 Boulogne Billancourt,
France.

Dr. P. F. Hewkin,
BBC Brown Boverie & Cie,
Aktiengesellschaft,
Postfach 10 16 80,
6900 Heidelberg 1,
Germany.

Mr. E. Johnston,
Haden Drysys International Ltd.,
Alliance Road,
London W2 0RA.

Dr. Robert B. Kelley,
Director, Robotics Research Center,
Kelley Hall,
University of Rhode Island,
Kingston, RI 02881,
U.S.A.

Dr. P. W. Kitchin,
Patscentre Benelux SA,
Avenue Rhine Astrid,
B-1430 Braine le Chateau,
Belgium.

Mr. M. Kohno,
Production Engineering Research
Laboratory,
Hitachi Limited,
292 Yoshida-Machi,
Totsuka-ku,
Yokohama 244,
Japan.

Dr. Arkady G. Makhlin,
Westinghouse Electric Corporation,
Electronic Systems Div.,
Box 160,
Pittsburgh, Pennsylvania 15230,
U.S.A.

Mr. Peter Martini,
Universitat Karlsruhe,
Institut fur Informatik III,
Postfach 6380,
Zirkel 2,
7500 Karlsruhe 1,
Germany.

Professor M. Nougaret,
Lab. d'Automatique de Grenoble-
E.N.S.I.E.G.,
B.P. 46,
38402 St. Martin D'Heres,
France.

Professor Alan Pugh,
Department of Electronic Engineering,
University of Hull,
Hull HU6 7RX.

Mr. Gordon I. Robertson,
President,
Control Automation Inc.,
PO Box 2304,
Princeton, New Jersey 08540,
U.S.A.

Mr. Lothar Rossol,
Vice President of R&D,
GMFanuc Robotics Corporation,
5600 New King Street,
Troy, Michigan 48098,
U.S.A.

Professor Arthur C. Sanderson,
Robotics Institute,
Carnegie-Mellon University,
Pittsburgh, PA 15213,
U.S.A.

Dr. P. Saraga,
Circuit and Vaccum Physics Division,
Philips Research Laboratories,
Cross Oak Lane,
Redhill,
Surrey RH1 5HA.

Mr. Philippe Villers,
President,
Automatix Inc.,
217 Middlesex Turnpike,
Burlington, MA 01803,
U.S.A.